Gas Explosion and Its Protection Technology in Process Industries

Zhirong Wang · Xingyan Cao

Gas Explosion and Its Protection Technology in Process Industries

 Springer

Zhirong Wang
College of Emergency Management
Nanjing Tech University
Nanjing, Jiangsu, China

Xingyan Cao
College of Safety Science and Engineering
Nanjing Tech University
Nanjing, Jiangsu, China

ISBN 978-981-96-3120-9   ISBN 978-981-96-3121-6 (eBook)
https://doi.org/10.1007/978-981-96-3121-6

This Springer imprint is published by the registered company Springer Nature Singapore Pte Ltd.
The registered company address is: 152 Beach Road, #21-01/04 Gateway East, Singapore 189721, Singapore

If disposing of this product, please recycle the paper.

Contents

Chapter 1
Introduction

Gas explosion accidents, which are one of the main forms of industrial disasters, frequently occur in the process of production, usage, storage, and transportation of combustible gases. The explosions of combustible gases cause significant losses to the safety of people's lives and property, severely hindering the development of the national economic and social progress. Effectively preventing, controlling and reducing the economic losses and casualties caused by gas explosions has become a key issue that needs to be solved. Due to the highly destructive and complex nature of gas explosion accidents, various countries have invested substantial financial resources into extensive research and have achieved considerable results.

Scholars around the world have carried out a lot of research on the propagation process of gas explosions in industrial confined spaces [1]. When flammable gas is ignited in a confined space, the flame propagates rapidly to the unburned gas [2]. Based on the rate of energy released during the explosion and the propagation velocity of the hot gas products, the explosion is classified into deflagration and detonation. The deflagration propagates at subsonic speeds, with propagation velocities typically ranging from a few meters to several hundred meters per second, while the detonation propagates at supersonic speeds of kilometers per second [3]. The deflagration creates a pressure wave in front of the flame, forming a 'two-wave, three-zone' (combusted zone–flame wave–preheated zone–pressure wave–unburned zone) structure [4]. At this time, the flame self-instability and turbulence caused by the walls of the confined space will accelerate the flame propagation [5]. The types, concentrations and environmental conditions of gases also affect the deflagration intensity and flame propagation characteristics [6]. Studies have shown that the higher the gas activity and the closer the concentration is to the stoichiometric concentration the faster the explosion releases energy and the greater the explosion overpressure and flame propagation velocity [7]. Properly arranged obstacles can lead to an increase in flame area and trigger turbulent combustion, which has a significant acceleration effect on the flame [8–10]. The continuous acceleration of flame propagation triggers the deflagration-to-detonation transition (DDT) phenomenon [11]. Currently, a hot

© The Author(s) 2025
Z. Wang and X. Cao, *Gas Explosion and Its Protection Technology in Process Industries*, https://doi.org/10.1007/978-981-96-3121-6_1

spot appears behind the shock wave or Mach bar, and the flame gradually catches up with the precursor shock wave and couples with it [12–14]. The DDT process generates a high overpressure, the value of which is even greater than the steady detonation pressure, and has a great destructive power [15]. However, its occurrence is more random and the process is rapid, and the critical mechanism and the intrinsic mechanism of its occurrence still need to be further revealed [16].

Connected vessels are a common type of structure in industry and their explosion mechanisms are more complex than those of a single vessel or pipe. It has been shown that the explosion flame in a connected vessel creates strong turbulence when it reaches the secondary vessel, causing a more violent explosion [17], resulting in a higher maximum overpressure and higher rate of pressure rise [18]. Composition structure, dimensions and ignition location affect the gas explosion characteristics of connected vessels [19]. Connected form of spherical vessel–pipe–spherical vessel will form two peaks of overpressure, while the connected vessel of spherical vessel–pipe form, the overpressure has a single peak, and the spherical vessel will form a large negative pressure inside [20]. Wang et al. [21] studied the effect of changes in the volume of cylindrical container and pipe size on the overpressure of methane–air mixtures, and found that the maximum rate of increase in overpressure decreases with the increase in the volume size of the container. Bartknecht et al. [22] found that when the volume of the container is the same, the length of connecting pipe does not have much impact on the explosion intensity, while the diameter of the pipe has a greater impact on the explosion intensity. You et al. [23] pointed out that when the pipe diameter constant, the larger the volume of primary vessel, the greater the initial velocity of the flame entering the pipe. The smaller the volume of secondary vessel, the stronger the resistance to the flame propagation, and the slower the flame reaches the secondary vessel.

To mitigate the risk of gas explosions in confined spaces, three primary methods are typically employed to lessen the impact of such events: explosion venting, explosion resistance, and explosion suppression. Explosive venting can protect the container unit by releasing the pressure wave, flame and unburned combusted media in the confined space to the external space [24]. Explosive venting shows significant differences at various combustible gas concentrations, venting areas, or rupture pressures. Higher concentration leads to a secondary explosion on the outside of the vent. The maximum secondary overpressure increases linearly with the maximum vessel pressure [25]. The maximum overpressure increased with bursting pressure of bursting film and decreased with venting area [26]. Explosion arrest is another important method of explosion protection. The goal is to inhibit the propagation of explosion flames resulting from flammable gases and vapors of flammable liquids within the pipe and equipment systems. A flame arrester is a commonly employed safety device in the industry for preventing explosions. It exhibits remarkable extinguishing capabilities on explosion flames and effectively halts the spread of such flames within pipe systems [27]. Within a flame arrester, the microchannels play a crucial role in absorbing substantial heat and reactive free radicals. The flame extinguishes gradually when the heat absorbed by the microchannel walls surpasses that released by combustion, or when there is not enough reactive radicals to sustain the chemical reaction process [28, 29]. Enhancing the quenching effect on gas explosion flames is achievable by augmenting the thickness and porosity of the flame arrester unit. Moreover, the level of porosity significantly impacts the effectiveness of flame quenching [30]. The wire mesh offers a level of blast resistance, influencing the structural properties of the flame front surface to hinder flame spread [31]. The geometric attributes of the wire mesh dictate critical parameters like quenching velocity and overpressure for successful blast resistance [32]. Enhancing blast resistance is achievable by increasing mesh size and the number of wire mesh layers [33]. Apart from flame arresters and wire mesh, other porous materials like foam ceramic pore structures can quench explosion flames and provide detonation resistance [34, 35]. The quenching efficacy of porous materials stems from mechanisms such as the wall effect, heat dissipation, and absorption of transverse waves [36]. By leveraging quenching principles in designing novel porous material structures and compositions, more resilient porous materials with enhanced explosion resistance can be engineered [37, 38].

Explosion suppression stands out as a highly effective approach for mitigating the intensity and destructive potential of explosions. Extensive scientific inquiry has delved into the efficacy of inhibitors in explosion suppression. Among these, ultra-fine water mist, renowned for its robust heat absorption capabilities and eco-friendly traits, emerges as a top-notch explosion suppressant. Research indicates that ultra-fine water mist excels in curbing overpressure, flame burning rates, flame temperatures, and flame structures, showcasing superior suppression performance [39–41]. The primary reasons behind the deceleration of flame burning rates by ultra-fine water mist include heat absorption, dilution, and chemical suppression, with the latter playing a relatively minor yet significant role [42]. With escalating water mist concentrations, there is a correlated decrease in overpressure, flame propagation rates, and flame temperatures. The efficacy of suppression is intricately tied to both the concentration of combustible gases and the dosage of applied water mist [43]. Even when encountering obstacles, augmenting the concentration of water mist can effectively diminish overpressure and flame temperatures, potentially leading to complete explosion suppression once a critical threshold is reached [44]. Notably, larger droplets in water mist, if not evaporated in time, can act as obstacles, inducing flame turbulence. This turbulence accelerates flame combustion rates, intensifying the explosion reaction. Furthermore, as particle size increases, the turbulence effect becomes more pronounced [45, 46]. To ensure efficient heat absorption, water mist particle sizes should ideally fall within the micron range. This size range enables effective heat absorption from the flame surface without significantly disrupting the flame front's flow field [47]. Introducing specific additives can enhance the explosion suppression efficacy of ultrafine water mist, leading to reduced flame propagation velocities, flame temperatures, and overpressures [48, 49]. Compared to pure ultrafine water mist, formulations with additives exhibit superior cooling abilities, heightened thermal radiation blocking capabilities, and lower concentrations of chain reaction radical substances (H, OH, and O). The inhibitory effect is notably enhanced with increasing concentrations of ultrafine water mist containing additives [50]. Additionally, the choice of additives significantly impacts the explosion suppression potential of ultrafine water mist, with KOH demonstrating superior suppression effects compared to NaOH and NaCl [51].

In conclusion, current research trends, both domestically and internationally, regarding the characteristics of explosions in confined spaces emphasize studying the mechanisms and laws governing the evolution of explosion dynamics. This includes developing accurate theoretical mathematical models to describe the explosion processes, utilizing fluid dynamics software or custom software for simulating explosions, uncovering fundamental rules of explosion processes, and applying research findings to analyze accident consequences, as well as for designing industrial equipment explosion prevention and safety measures. Considering these trends, the authors have leveraged the resources of the Key Laboratory of Chemical Process Safety under the Ministry of Emergency Management to systematically delve into gas-phase explosions in confined spaces. This book represents a consolidation of the primary theoretical approaches and key findings related to gas-phase explosions in confined spaces, drawing from advancements in both domestic and international research. The aim is to equip readers with a more systematic, comprehensive, and profound grasp of gas-phase explosions in confined spaces and their protective technologies. It is hoped that this book will not only enhance understanding but also offer guidance for future research endeavors in this critical area.

References

1. Shao, W., Bi, M. S., Ding, X. W., et al. (2002). Discussion on the flame acceleration mechanism of combustible gas cloud explosion. *Chemical Engineering and Machinery, 29*(2), 113–115.
2. Bi, M. S., Yin, W. H., & Ding, X. W. (2003). Numerical simulation of the explosion process of non-ideal explosion sources in a closed container. *Chemical Industry and Engineering Technology, 24*(3), 1–3.
3. Oran, E. S., & Williams, F. A. (2012). The physics, chemistry and dynamics of explosions. *Philosophical Transactions of the Royal Society A—Mathematical Physical and Engineering Sciences, 370*(1960), 534–543.
4. Hu, T. Z. (2008). *Numerical simulation study on the propagation law of gas explosions* [Master's thesis]. China University of Mining and Technology, Beijing.
5. Yang, P., Wang, T., Sheng, Y., et al. (2024). Recent advances in hydrogen process safety: Deflagration behaviors and explosion mitigation strategies. *Process Safety and Environmental Protection, 188*, 303–316.
6. Oran, E. S., Chamberlain, G., & Pekalski, A. (2020). Mechanisms and occurrence of detonations in vapor cloud explosions. *Progress in Energy and Combustion Science, 77*, 100804.
7. Mercx, W. P. M., & Van den Berg, A. C. (1997). The explosion blast prediction model in the revised CPR 14E (yellow book). *Process Safety Progress, 16*(3), 152–159.
8. Lin, B. Q., & Gui, X. H. (2001). Simulation study on the flame propagation law in gas explosions. *Journal of China University of Mining and Technology, 31*(1), 6–9.
9. Lin, B. Q., & Gui, X. H. (2002). Measurement of flame thickness and numerical simulation analysis of temperature field in gas explosions. *Experimental Mechanics, 17*(2), 227–233.
10. Zheng, K. (2017). *Study on the propagation characteristics of premixed flames in hydrogen/ methane mixed fuel deflagration in pipelines* [Doctoral dissertation]. Chongqing University.

11. Fairweather, M., Hargrave, G. K., Ibrahim, S. S., et al. (1999). Studies of premixed flame propagation in explosion tubes. *Combustion and Flame, 116*(4), 504–518.
12. Wang, Z. R. (2005). *Study on gas explosion propagation and its dynamic process in confined spaces* [Doctoral dissertation]. Nanjing Tech University.
13. Saif, M., Wang, W., Pekalski, A., et al. (2017). Chapman–Jouguet deflagrations and their transition to detonation. *Proceedings of the Combustion Institute, 36*(2), 2771–2779.
14. Kessler, D. A., Gamezo, R. N., & Oran, R. S. (2010). Simulations of flame acceleration and deflagration-to-detonation transitions in methane–air systems. *Combustion and Flame, 157*(11), 2063–2077.
15. Khokhlov, A., & Oran, E. (2013). Adaptive mesh numerical simulation of deflagration-to-detonation transition: The dynamics of hot spots. In *Proceedings of the 30th Fluid Dynamics Conference*.
16. Qian, J., Gao, Y., Liu, Z., et al. (2024). The effect of concentration gradient on overpressure hazards and flame behavior of gas explosion in a vessel-duct connected device. *Fuel, 371*, 131901.
17. Phylaktou, H. N., & Andrews, G. E. (1993). Gas explosions in linked vessels. *Journal of Loss Prevention in the Process Industries, 6*(1), 15–19.
18. Maremonti, M., Russo, G., Salzano, E., et al. (1999). Numerical simulation of gas explosions in linked vessels. *Journal of Loss Prevention in the Process Industries, 12*(3), 189–194.
19. Cui, Y. Q. (2013). *Study on the influence of structure and size on methane-air premixed gas explosions in containers and pipelines* [Master's thesis]. Nanjing Tech University.
20. Wang, Z. R., Jiang, J. C., & Zhou, C. (2011). Experimental study on gas explosion characteristics in connected devices. *Explosions and Shock Waves, 31*(1), 6.
21. Wang, Z. R., Sun, P. P., Tang, Z. H., et al. (2021). Size effect on the explosion of methane-air mixtures in closed containers. *Journal of Safety Science and Technology, 31*(1), 60–66.
22. Bartknecht, W. (1981). *Explosion course prevention protection*. Springer-Verlag.
23. You, M. W., Yu, Y., Jiang, J. C., et al. (2012). Explosion of premixed gases in connected containers under different pipe length conditions. *Combustion Science and Technology, 18*(03), 256–259.
24. Zhang, W., Jiang, J. C., Wang, Z. R., et al. (2018). Study on the pressure characteristics of single/double venting in connected containers. *Journal of Safety Science and Technology, 28*(8), 43.
25. Xing, H. D. (2021). *Study on the mechanism and flow field characteristics of methane gas venting* [Master's thesis]. Nanjing University of Science and Technology.
26. Lu, Y., Wang, Z., Cao, X., et al. (2021). Interaction mechanism of wire mesh inhibition and ducted venting on methane explosion. *Fuel, 304*, 121343.
27. Lv, X., Yu, J., Hou, Y., et al. (2022). The quenching and attenuation of hydrogen-air detonation after passing across capillaries. *Fuel, 324*, 124535.
28. Kim, K. T., Lee, D. H., & Kwon, S. (2006). Effects of thermal and chemical surface-flame interaction on flame quenching. *Combustion and Flame, 146*(1–2), 19–28.
29. Mahuthannan, A. M., Damazo, J. S., Kwon, E., et al. (2019). Effect of propagation speed on the quenching of methane, propane, and ethylene premixed flames between parallel flat plates. *Fuel, 256*, 115870.
30. Chendi, L., Xingyan, C., Zhirong, W., et al. (2022). Research on quenching performance and multi-factor influence law of hydrogen crimped-ribbon flame arrester using response surface methodology. *Fuel, 326*, 124911.
31. Xianfeng, C., Zhao, Q., Huaming, D., et al. (2018). Effect of metal mesh on the flame propagation characteristics of wheat starch dust. *Journal of Loss Prevention in the Process Industries, 55*, 107–112.
32. Cui, Y. Y., Wang, Z. R., Zhou, K. B., et al. (2016). Effect of wire mesh on double-suppression of CH_4/air mixture explosions in a spherical vessel connected to pipelines. *Journal of Loss Prevention in the Process Industries, 45*, 69–77.
33. Yu, J. L., Cai, T., & Li, Y. (2008). Experimental study on the effect of mesh structure on the quenching of explosive gases. *Combustion Science and Technology, 14*(2), 13–16.

34. Guo, C. M., & Chen, Z. G. (2000). Experimental study on the effect of porous steel plates on the propagation of detonation waves. *Experimental Mechanics, 15*(4), 400–407.
35. Guo, C. M., & Li, J. (2000). Experimental study on sound absorption of detonation waves in damped pipelines. *Explosions and Shock Waves, 20*(4), 289–295.
36. Nie, B. S., He, X. Q., Zhang, J. F., et al. (2008). Experimental study and mechanism of foam ceramics on the gas explosion process. *Journal of Coal Science and Engineering, 33*(8), 903–907.
37. Yuan, B., He, Y., Chen, X., et al. (2022). Flame and shock wave evolution characteristics of methane explosion in a closed horizontal pipeline filled with a three-dimensional mesh porous material. *Energy, 260*, 125137.
38. Li, Y., Wang, M., Zhang, G., et al. (2024). Isolation effectiveness of combined rigid-flexible porous materials against methane/hydrogen explosion. *International Journal of Hydrogen Energy, 58*, 93–104.
39. Qin, W. Q., Wang, X. S., Gu, R., & Xu, H. L. (2012). Explosion pressure and rate of increase under ultra-fine water mist action. *Combustion Science and Technology, 18*(1), 90–95.
40. Luo, Z. M. (2008). Experiment on controlling gas explosion with activated steam. *Geology and Exploration of Coalfields, 36*(3), 23–30.
41. Tang, J. J. (2009). *Experimental and numerical study on the suppression of gas explosion with fine water mist* [Doctoral dissertation]. Xi'an University of Science and Technology.
42. Akira, Y., Toichiro, O., Wataru, E., & Hiroyoshi, N. (2015). Experimental and numerical investigation of flame speed retardation by water mist. *Combustion and Flame, 162*, 1772–1777.
43. Gu, R., Wang, X. S., & Xu, H. L. (2010). Experimental study on suppression of methane explosion with ultra-fine water mist. *Fire Safety Science, 19*(2), 546–553.
44. Qin, W. Q. (2011). *Experimental study on the suppression of methane explosion with ultra-fine water mist containing obstacles* [Doctoral dissertation]. University of Science and Technology of China.
45. Thomas, G. O. (2000). On the conditions required for explosion mitigation by water sprays. *Process Safety and Environmental Protection, 78*(5), 339–354.
46. Gieras, M. (2008). Flame acceleration due to water droplets action. *Process Safety and Environmental Protection, 21*(4), 472–477.
47. Cao, X. Y., Ren, J. J., Zhou, Y. H., et al. (2016). Mechanism analysis of enhancement and suppression of methane/air explosion by ultrafine water mist. *Journal of Coal Science and Engineering, 41*(7), 1711–1719.
48. Zheng, R., Bray, K. N. C., & Rogg, B. (1997). Effect of sprays of water and NaCl-water solution on the extinction of laminar premixed methane-air counterflow flames. *Combustion Science and Technology, 126*, 389–401.
49. An, A. (2011). *Experimental study on the suppression of pipeline gas explosion with fine water mist* [Doctoral dissertation]. Henan Polytechnic University.
50. Cao, X. Y., Ren, J. J., Zhou, Y. H., et al. (2015). Suppression of methane/air explosion by ultrafine water mist containing sodium chloride additive. *Journal of Hazardous Materials, 285*, 311–318.
51. Chelliah, H. K., Lazzarini, A. K., Wanigarathne, P. C., et al. (2002). Inhibition of premixed and non-premixed flames with fine droplets of water and solutions. *Proceedings of the Combustion Institute, 29*, 369–376.

Chapter 2
Gas Explosion Characteristics in the Confined Spaces

2.1 Gas Explosion Characteristics in the Single Vessel

2.1.1 Experimental Apparatus and Methods

2.1.1.1 Experimental Apparatus

Figure 2.1 illustrates a schematic diagram of the experimental apparatus for studying the syngas explosion characteristics [1]. It mainly consisted of a visualization explosion vessel, a gas supply system, an ignition system, a high-speed camera system, a high-frequency pressure acquisition system, a procedure control and data acquisition system, and the obstacle board.

The visualization explosion vessel was made of 304 stainless steel (150 mm × 150 mm × 910 mm). To capture the flame propagation characteristics after passing through the obstacle, two tempered glasses were installed respectively in the front and back sides of the vessel. The gas premixed system consisted of a premixed tank and a gas source, and a certain volume concentration is prepared by the partial pressure method. A high-voltage discharge method (voltage (U) = 8 kV) with a 5 mm electrode gap (height distance from bottom end (h) = 100 mm) was adopted to ignite the premixed gas. A high-speed camera (acquisition frequency (f) = 4000 fps; resolution (h × t) = 1280 × 1024) is used to capture the flame propagation process after passing through the obstacle. A pressure transmitter (measurement accuracy: 0.25% FS; range: 2.50 MPa) with 50 kHz response frequency was installed in the middle of the vessel to acquire the pressure data. A data acquisition card with 200 kHz frequency (USB287X) was adopted to realize the program control and data acquisition.

© The Author(s) 2025
Z. Wang and X. Cao, *Gas Explosion and Its Protection Technology in Process Industries*, https://doi.org/10.1007/978-981-96-3121-6_2

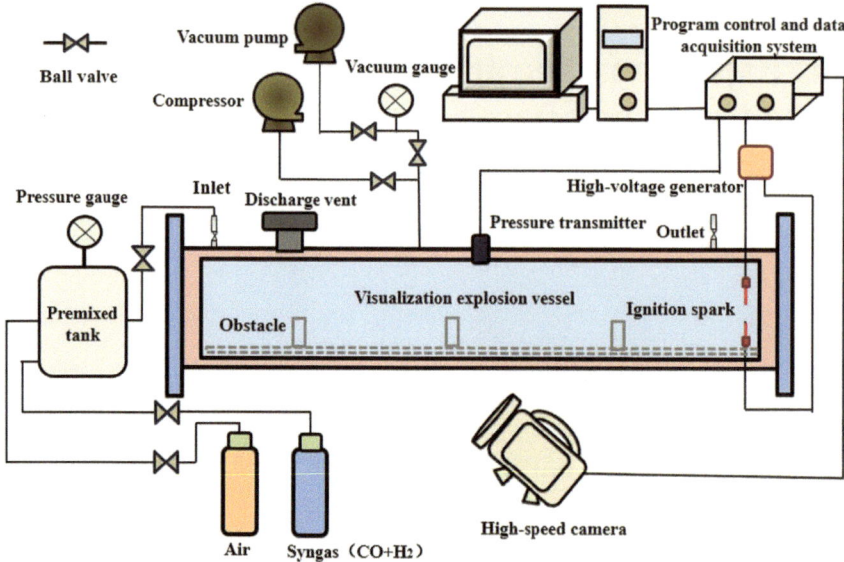

Fig. 2.1 Schematic diagram of the visualization experimental apparatus

2.1.1.2 Experimental Methods

Before the experiment, the vacuum degree of explosion vessel was pumped to 0.09 MPa. The premixed gas with a certain concentration was introduced into the explosion vessel and laid aside for 10 min. Before ignition, the initial pressure inside the vessel was 0.10 MPa and the volume concentration was determined. After ignition, the flame accelerated propagation from right to left and bypassed the obstacle. Each condition was repeated 3–4 times to guarantee the experimental accuracy.

2.1.1.3 Experimental Conditions

Considering the axial length and cross section of explosion vessel, the obstacles (900 mm × 150 mm × 10 mm) of three blocking rates (ε = 0, 20 and 33%) and four obstacle numbers (n = 0, 1, 2 and 3) were selected. The thicknesses of the obstacle and its bottom board were 10 mm and 1 mm respectively and the material was 304 stainless steel. The distance between the first obstacle and ignition source was 170 mm and the distance between adjacent obstacles was 200 mm, respectively. Three syngas concentrations (CO/H_2 = 1:1; c = 15, 20 and 25% (the equivalence ratio (Φ) = 0.51, 0.68 and 0.84)) were adopted to carried out the experiment. The effect mechanism of obstacle was studied by the changings of flame propagation characteristics and explosion parameters under different working conditions.

2.1.2 Effect of Gas Concentration

Figure 2.2 shows the variation history of flame front position with time under 15, 20 and 25% syngas concentrations ($n = 2$; $\varepsilon = 33\%$). It clears that the time for the flame to reach the vessel end was decreased successively with increasing syngas concentration (from 266.00 to 39.50 ms). The flame propagation velocity was calculated according to the variation of flame front position with time, as shown in Fig. 2.3. The interval time of flame front position change was reduced as much as possible ($\Delta t = 1$ ms), so that the average velocity was close to the instantaneous velocity of flame propagation at a certain moment. The flame propagation velocity was increased obviously with the increase of syngas concentration. Especially the three peaks of velocity history were increased significantly, and their appearance moments and the interval time between adjacent peak values were decreased in turn. And the v_{max} was increased by 88.90% and the interval time was reduced from 26.75 ms and 30.11 ms ($c = 15\%$) to 4.26 ms and 3.62 ms ($c = 25\%$), respectively. It was because that the chemical reaction rate was gradually enhanced with increasing syngas concentration under lean-fuel concentration. Besides, the flame turbulence intensity was enhanced under the action of obstacle, which resulted in the increase of flame propagation velocity. Compared with the high syngas concentration, the flame propagation of the low syngas concentration was relatively slow, and the intensities of combustion wave and reflected wave were relatively weak. The duration of both interactions was increased continuously, further resulting in a gradual increase in the duration of the flame oscillation.

Fig. 2.2 The variation history of flame front position with time under 15, 20 and 25% syngas concentrations ($n = 2$; $\varepsilon = 33\%$)

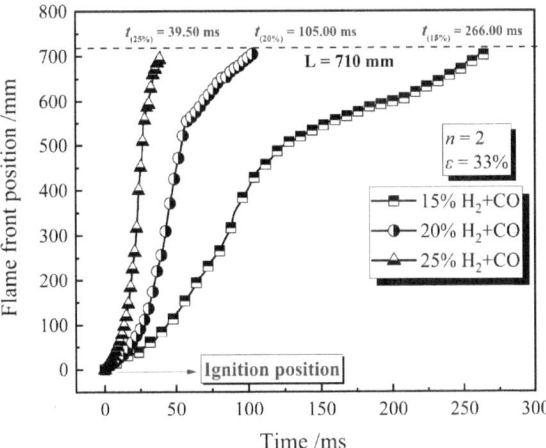

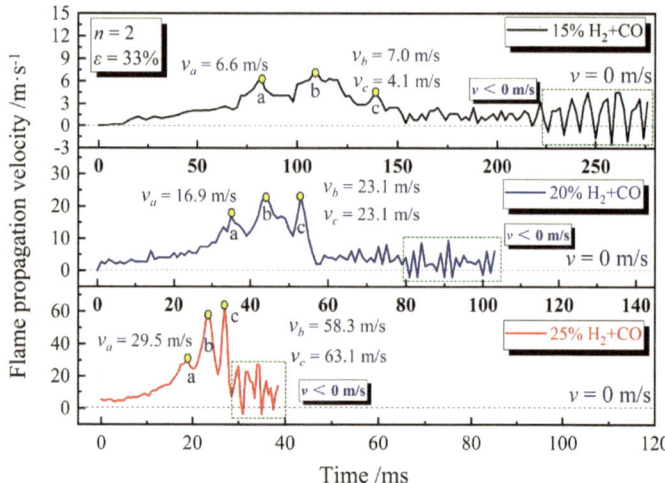

Fig. 2.3 Effect of syngas concentration on flame propagation velocity ($n = 2$; $\varepsilon = 33\%$)

2.1.3 Effect of Obstacle Number

Figure 2.4 shows the effect of obstacle number on the flame propagation velocity history ($n = 0$, 1, 2 and 3; $c = 25\%$; $\varepsilon = 33\%$). It clears that the flame propagation velocity was increased significantly and the propagation time inside the vessel was decreased successively with increasing obstacle number (from 68.50 to 33.00 ms). Meanwhile, the number of the velocity history peak was related to the obstacle number. The propagation velocity appeared a change process of the increase firstly, then decrease and again increase as the flame bypassed the obstacle. Hence, the peak number of velocity history was more one than the obstacle number. The characteristics of flame propagation was also studied in Luo's experiment [2]. However, the relationship between the velocity history peak and obstacles was not mentioned. As can be seen from Fig. 2.4, the peak of velocity history was increased continuously with increasing obstacle number. And the v_{max} was increased by 60.59%. The peak variation tended to be stable under three obstacles, but the fourth acceleration propagation process still existed. Moreover, the extent of flame "reverse oscillation" decreased with the increase of obstacle number compared with one obstacle. Especially the flame "reverse oscillation" disappeared under three obstacles. It was because that the interaction intensity of combustion and compression waves was enhanced and gradually played a dominant role due to the enhancement of flame turbulence disturbance with the obstacle number increased. And the reflected wave intensity was relatively weakened, further resulting in the reduction of flame "reverse oscillation" phenomenon.

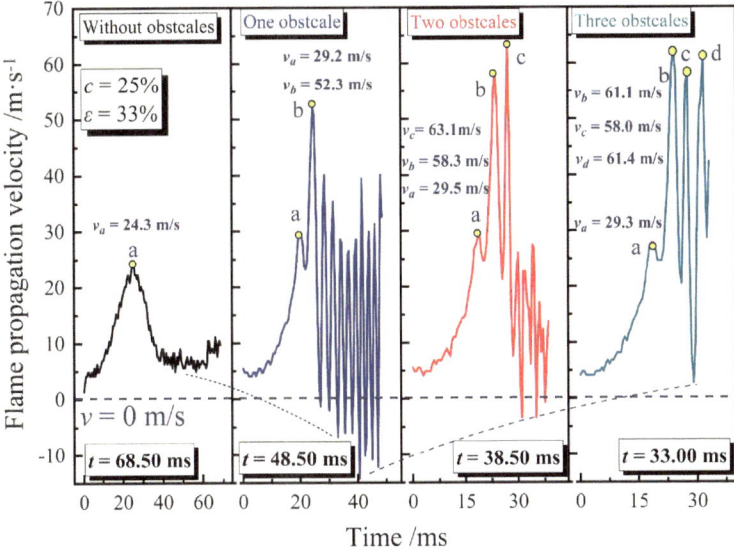

Fig. 2.4 Effect of obstacle number on the flame propagation velocity (c = 25%; ε = 33%)

2.1.4   *Effect of Blocking Rate*

Figure 2.5 presents the effect of obstacle blocking rate on the flame propagation velocity (ε = 0, 20 and 33%; c = 25%; n = 2).

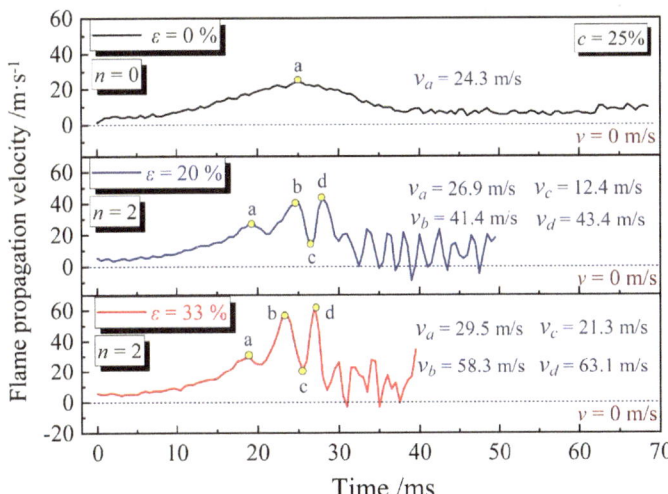

Fig. 2.5 Effect of obstacle blocking rate on flame propagation velocity (c = 25%; n = 2)

The flame propagation velocity was increased, and the time required for the flame to reach the vessel end was decreased as the blocking rate increased (from 68.50 to 39.50 ms). Meanwhile, the corresponding velocity histories all showed three peaks as two obstacles existed, which indicated that the flame all experienced three times acceleration propagation process. The corresponding velocity peak was also increased continuously as the blocking rate increased, and their appearance moment showed a decreasing trend. The v_{max} was increased by 61.65% and the interval time was reduced from 5.49 ms and 3.42 ms ($\varepsilon = 20\%$) to 4.45 ms and 3.85 ms ($\varepsilon = 30\%$), respectively. It was because that the extent of stretch and twist was increased as the flame bypassed the high blocking rate, and the turbulence disturbance of flow field was also enhanced significantly.

2.2 Gas Explosion Characteristics in the Connected Vessel

2.2.1 Experimental Apparatus and Methods

2.2.1.1 Experimental Apparatus

As is illustrated in Fig. 2.6, the experimental system was made up of explosion system, data acquisition system and auto gas distribution system.

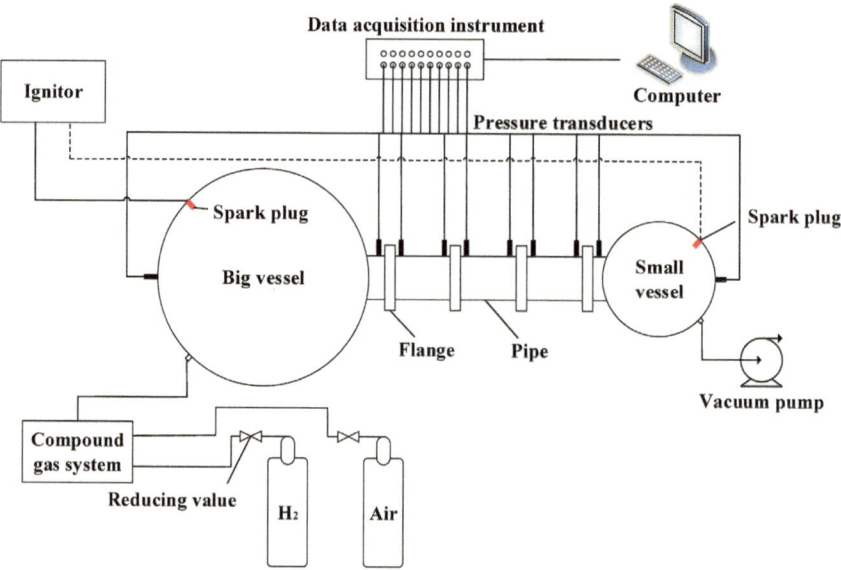

Fig. 2.6 Schematic diagram of the explosion apparatus for linked vessels

The explosion system was made up of explosion apparatus and ignition apparatus. The explosion apparatus was made up of a big spherical vessel, pipes, and a smaller spherical vessel. The big spherical vessel is 600 mm in diameter and 113 L in volume. The smaller spherical vessel is 350 mm in diameter and 22 L in volume. Each cylindrical pipe was 2000 mm in length and 60 mm in internal diameter. The wall thicknesses of the pipe, the big vessel, and the smaller vessel are 15 mm, 22 mm, and 16 mm respectively. All these three components were connected by flanges. Nozzles in the spherical vessels were used to install pressure transmitters, vacuum manometers, spark plugs, a gas inlet, and a gas outlet. An KTD-A ignition apparatus was used as a capacitive electrical ignition source for the high-voltage spark plugs. This device was simple and accurate to adjust. In these experiments, the ignition position was in the center of vessel.

The data acquisition system was made up of pressure transducer and data acquisition device. Overpressure was obtained by high frequency pressure transducer named HM 90-H3-2. The pressure transducer range was 0–10 MPa. The frequency response was 200 kHz. The output signal was 0–5 V DC, and the measuring accuracy was ± 0.25% FS. The pressure transducer was placed on the wall of each spherical vessel. The resolution of the data acquisition device was 24-bit, and the sampling rate was 200 kHz. A DEWE-43 multi-channel data acquisition device was adopted to collect synchronous data. The data analysis software named DEWE Soft X2 was used to process the data.

The auto gas distribution system was made up of distribution system and a vacuum pump. A RCSC2000-B distribution system, which was made up of distribution box, control box, computer, hydrogen gas cylinder and oxygen gas cylinder, was used to get the mixture of hydrogen and air of the desired concentration. The process of gas distribution was auto finished by connecting to computer. A software named computer automatic gas distribution was used to control this process. A 2X-8 GA vacuum pump was used to replace the waste gas in the linked vessels with fresh air. Besides, the fastest pumping rate can achieve 0.008 m^3/min, which was enough for the experiment.

2.2.1.2 Experimental Methods

The experiments were carried out at normal temperature and atmospheric pressure, and the ignition energy was 5 J. Before each experiment, the vacuum pump was used to get the air pressure in the linked vessels at − 0.09 MPa, then to aerate hydrogen–air mixture as required. In order to establish the repeatability of experiments, each experiment was repeated three times.

2.2.1.3 Experimental Conditions

The experimental scheme is shown in Table 2.1.

Table 2.1 Experimental scheme for the influential factors on Hz-air explosion in linked vessels

No.	Connection mode	Ignition position	Initial pressure (MPa)	Position of pressure transducer
1	Big vessel + 2.5 m pipe + small vessel	Small vessel	0	On the wall of each vessel
2	Big vessel + 2.5 m pipe + small vessel	Big vessel		
3	Big vessel + 4.5 m pipe + small vessel	Small vessel		
4	Big vessel + 4.5 m pipe + small vessel	Big vessel		
5	Big vessel + 6.5 m pipe + small vessel	Small vessel		
6	Big vessel + 6.5 m pipe + small vessel	Big vessel		
7	Big vessel + 2.5 m pipe + small vessel	Small vessel	− 0.02	
8			− 0.01	
9			0.01	
10			0.02	
11			0.03	

2.2.2 Effect of Ignition Position

Due to different flame propagation corresponding to different hydrogen explosion mechanism, the expanding direction of hydrogen was quietly different, which led to a different diffusion rate. Meanwhile, the explosive limit of hydrogen will be affected. The experiment was carried out at the condition of normal temperature and pressure. The connection mode was big vessel–pipe–small vessel. The flame propagation has two options, one was from the small vessel to the big vessel and the other was from the big vessel to the small vessel. Also, the ignition position was located in the center of vessels. The recorded overpressure in big and small vessels are shown in Figs. 2.7, 2.8, 2.9 and 2.10. These figures showed that, in agreement with the result in [2], ignition position plays a great role on gas explosion in linked vessels. Figures 2.9 and 2.10 also showed that the maximum overpressure decreases with the decreasing of volume ratio. Different ignition positions had the same effect on gas explosion with different volume ratios.

When the ignition position was located in the center of small vessel, the flame propagation was from small vessel to big vessel. The overpressure in small vessel was 0.78 MPa. However, when the flame propagate to big vessel, the overpressure in big vessel was only 0.55 MPa. When the ignition position was in the center of big vessel, the flame propagation was changed from big vessel to small vessel. The overpressures in both vessels were increased significantly. The overpressure in big vessel and small vessel were 0.70 MPa and 1.20 MPa, respectively.

Fig. 2.7 Overpressure in the primary vessel connected with single pipe (ignited in both vessels)

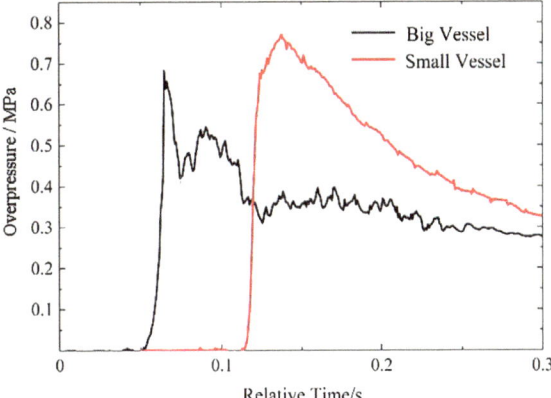

Fig. 2.8 Overpressure in the secondary vessel connected with single pipe (ignited in both vessels)

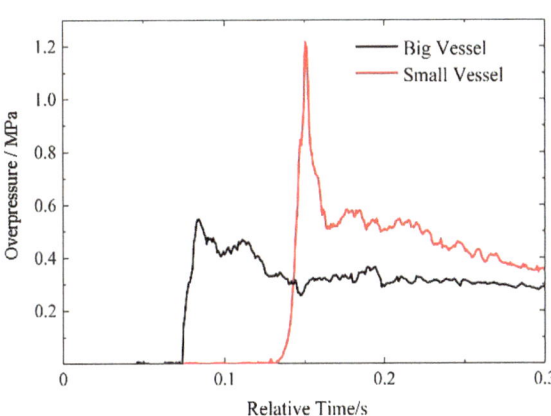

Fig. 2.9 Overpressure in the primary vessel and secondary vessel connected with single pipe (ignited in big vessel)

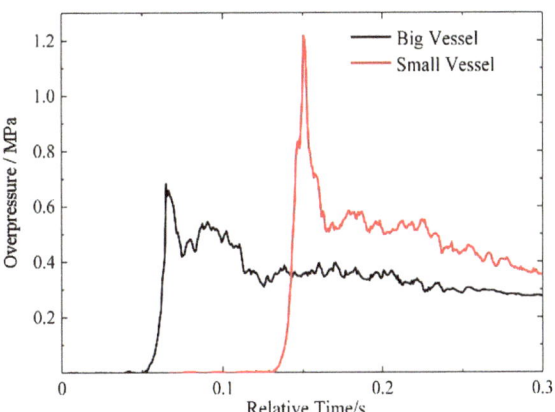

Fig. 2.10 Overpressure in the primary vessel and secondary vessel connected with single pipe (ignited in small vessel)

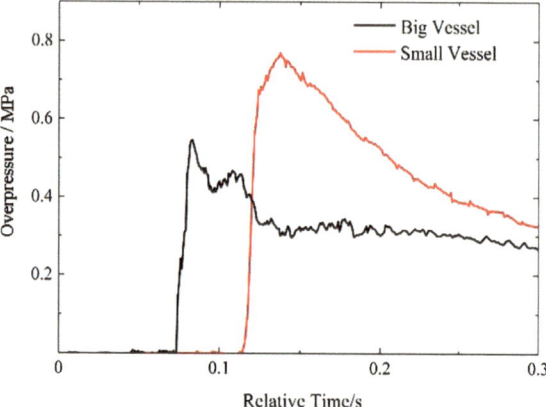

When ignited in the big vessel, the explosion flame first spread from big vessel to pipe. The laminar flow of flame become unstable and finally the turbulent flow was formed. During the process of laminar flow turns to turbulent flow, the speed of flame increases rapidly. Besides, the overpressure rising rate has a sharp rise in small vessel [3]. On the contrary, the maximum hydrogen overpressure in big vessel changes slightly. When the flame flows through the pipe, the speed of blast wave was much larger than the speed of flame. Thus, the blast wave will first spread to small vessel, then meets the wall and reflects to the pipe. The blast wave will meet the flame in the pipe, which calls stack effect. After stack, the flame continue spread to small vessel, the speed of flame increases rapidly and turns to turbulence combustion gradually. Meanwhile, because of the accumulation effect of overpressure in small vessel, the overpressure increased severely and finally came to the maximum overpressure. The hydrogen in big vessel also reacted quickly, and the maximum overpressure in big vessel achieved in a very short time. As a result, the maximum overpressures in big vessel and in small vessel achieved at the same time. Although the hydrogen in linked vessels reacted completely, overpressure wave oscillation exists in linked vessels. Then overpressure wave oscillation weakens because of energy loss. At last, overpressure wave oscillation disappears.

When the ignition position was located in the center of small vessel, the maximum hydrogen overpressure in big vessel had a small decline. Due to the speed of blast wave was larger than the flame, the reflect phenomenon will occur as the blast wave spread into big vessel. For the volume of big vessel was larger than the small one, the blast wave loses a lot of energy and the reflect wave become stronger. Thus, when the reflect wave meets the flame wave, the retardation become larger. Then, when the hydrogen flame spread to the large vessel, the hydrogen explosion intensity was weakened.

2.2.3   Effect of Initial Pressure

The experiment apparatus was linked vessels with one section pipe, and the ignition position was located in the center of small vessel. Figures 2.11 and 2.12 show the Overpressures in both vessels in linked vessels with one section pipe under different initial pressure. It's easy to find that although the pressure change trend in both vessels are not same, the overpressures in big and small vessel are increasing with the increase of initial pressure. When the initial pressure was lower than 0.02 MPa, the overpressure rising rate in small vessel only had a small increase with the increase of initial pressure. When the initial pressure was 0.03 MPa, the overpressure had a severe increase in small vessel. Compared with the initial pressure was 0.02 MPa, the overpressure almost has a 0.50 MPa changes. The overpressure in big vessel increases with the increase of initial pressure. The overpressure in big vessel reached to 1.13 MPa when the initial pressure become to 0.03 MPa. With the increasing of initial pressure, the time reached to the maximum overpressure has increased.

Fig. 2.11 Overpressure in small vessel with different initial pressure when ignited in small vessel

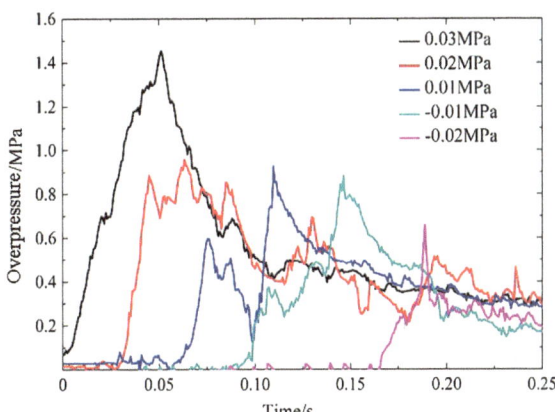

Fig. 2.12 Overpressure in big vessel with different initial pressure when ignited in small vessel

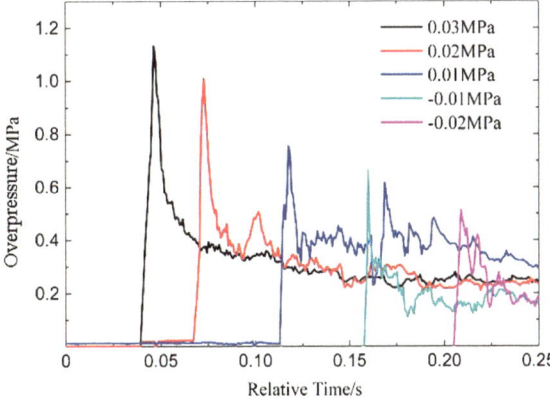

The collision theory considers that only highly reactive molecules collide can react together. Effective collision meant the new chemical bonds will form after the old chemical bonds separates in the unit number of collisions. A higher pressure led to a more molecular collision. Thus, even the initial pressure changed slightly, the hydrogen overpressure in the apparatus will have a change.

The gas content in small vessel will increase a lot with the increase of even 0.01 MPa initial pressure. Hence, when the explosion occurs, the released chemical energy from hydrogen molecules was higher [4]. With the increasing of initial pressure, the maximum overpressure and overpressure shock in both vessels became more severe. Besides, the higher initial pressure makes the small vessel has a part of unburned gas after ignited. While the flame or overpressure wave rebound from the big vessel, the turbulence explosion will occur. The overpressure and overpressure rising rate are both increase with the increasing of even 0.01 MPa initial pressure.

2.2.4 Effect of Pipe Length

Figure 2.13 shows the effect of pipe length on H_2-air explosion in linked vessels under the ignition position is in big vessel. With the increase of pipe length, the overpressures in both vessels were increasing. That is very dangerous for the device.

When ignited in the big vessel, the explosion flame first spread from big vessel to pipe. The laminar flow of flame become unstable and finally the turbulent flow was formed. During the process of laminar flow turns to turbulent flow, the speed of flame increases rapidly. Besides, the overpressure rising rate has a sharp rise.

Figure 2.14 shows the effect of pipe length on H_2-air explosion in linked vessels under the ignition position is in small vessel. The maximum overpressure in secondary vessel increased with the increasing of pipe length. However, the maximum overpressure in small vessel increased first then decreased. Due to the

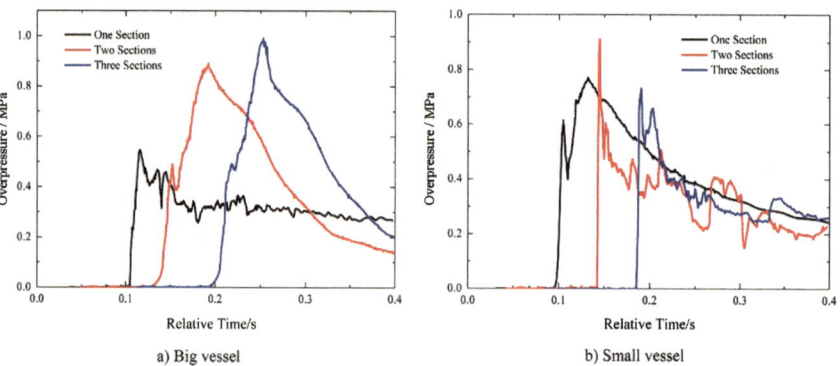

a) Big vessel   b) Small vessel

Fig. 2.13 Overpressures in linked vessels with different pipe length when ignited in big vessel

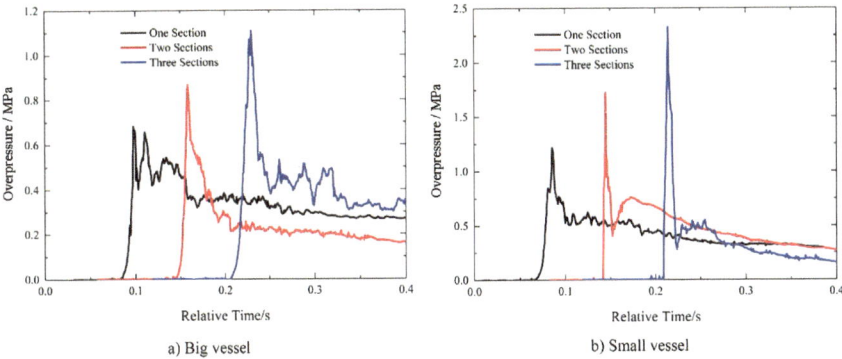

Fig. 2.14 Overpressures in linked vessels with different pipe length when ignited in small vessel

increasing of pipe length, the collision frequency between the flame, or overpressure wave and the pipe wall increased a lot. Mean-time, the energy dissipation also increased obviously. Thus, the maximum overpressure in small vessel decreased when the three-section pipe is connected.

References

1. Cao, X., Ren, J., Bi, M., et al. (2016). Experimental research on the characteristics of methane/air explosion affected by ultrafine water mist. *Journal of Hazardous Materials, 324,* 489–497.
2. Luo, Z., Kang, X., Wang, T., et al. (2021). Effects of an obstacle on the deflagration behavior of premixed liquefied petroleum gas-air mixtures in a closed duct. *Energy, 234,* 121291.
3. Canu, P., Rota, R., Carra, S., & Morbidelli, M. (1990). Vented gas deflagrations: A detailed mathematical model tuned on a large set of experimental data. *Combustion and Flame, 80*(1), 49–64.
4. Edwards, K. L., & Norris, M. J. (1999). Materials and constructions used in devices to prevent the spread of flames in pipelines and vessels. *Materials and Design, 20*(5), 245–252.

Chapter 3
Structure Effect of Connected Vessels on Gas Explosion Characteristics

3.1 Experimental Apparatus and Procedures

Figure 3.1 illustrates a schematic diagram of the experimental apparatus for studying the effect of connected vessels structure on methane explosion characteristics. The apparatus mainly consists of a primary vessel, a secondary vessel, a bifurcated pipe (including four connection forms), an ignition system, a high frequency pressure acquisition system, a gas premix system, a program control and data acquisition system.

Two kinds of spherical vessels (V = 22 and 113 L) and four kinds of cylindrical vessels (V = 11, 22, 55 and 113 L) were adopted as the connected vessel structure to carry out experimental research. The volume and shape of primary and secondary vessels were introduced in the corresponding section. The gas premixed system consisted of a gas flowmeter, a gas cylinder, a vacuum pump and a distribution controller. A high-energy ignition device (KTD-A) with 5 J energy was adopted to ignite premixed gas. Four pressure transmitters (measurement accuracy: 0.25% FS; range: 5 MPa) with 200 kHz response frequency were installed respectively in the primary/secondary vessel and the front and back positions of bifurcation pipe to acquire the pressure data of corresponding position, as shown in Fig. 3.1. A data acquisition card (DeweSoft TM) was used to realize the program control and data acquisition.

To research the effect of connected vessels structure on methane explosion (purity: 99.99%), four kinds of connection forms were selected. The connection between the vessel (spherical and cylindrical vessels) and bifurcated pipe is shown in Fig. 3.2. It consisted of a L-type connected vessels (Fig. 3.2a), two T-type connected vessels (Fig. 3.2b, c), and a Y-type connected vessels (Fig. 3.2d). The L-type connected vessels (abbreviated as S-T system) was used to research the effects of bending angle ($0° < \theta \leq 180°$), bending position (A, B, C and D) and vessel structure (vessel shape and volume). The T-type connected vessels were divided into T-L-R (i.e., the vessels connected at the left and right ends of pipe) and T-L-B (i.e., the vessels

© The Author(s) 2025
Z. Wang and X. Cao, *Gas Explosion and Its Protection Technology in Process Industries*, https://doi.org/10.1007/978-981-96-3121-6_3

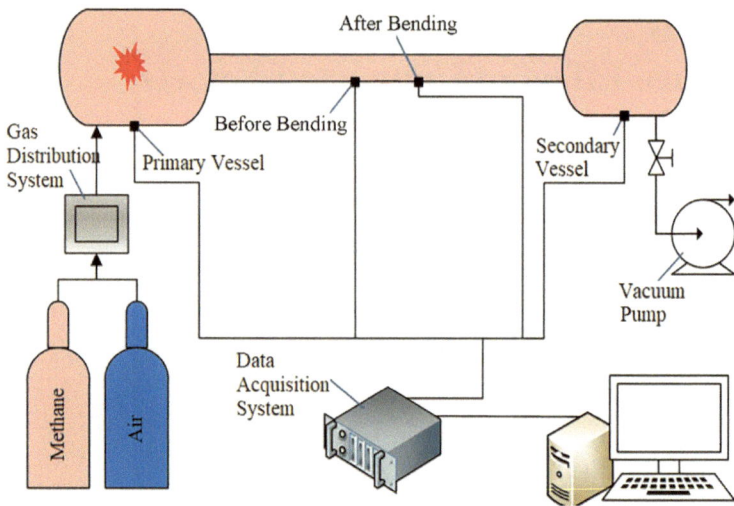

Fig. 3.1 Schematic diagram of the experimental apparatus

connected at the left and bottom ends of pipe) two connection types. The Y-type connected vessels (abbreviated as Y-L-R system) were the connection of two vessels forming a certain angle ($\theta = 150°$). The specifics of experimental scheme are listed in Tables 3.1 and 3.2. The inner diameter of main pipe which connected the two vessels was 59 mm and connected with two vessels by the flange, and the pipe end without connecting with any vessel was called bifurcated pipe. The ignition position was located at the geometric center of primary vessel.

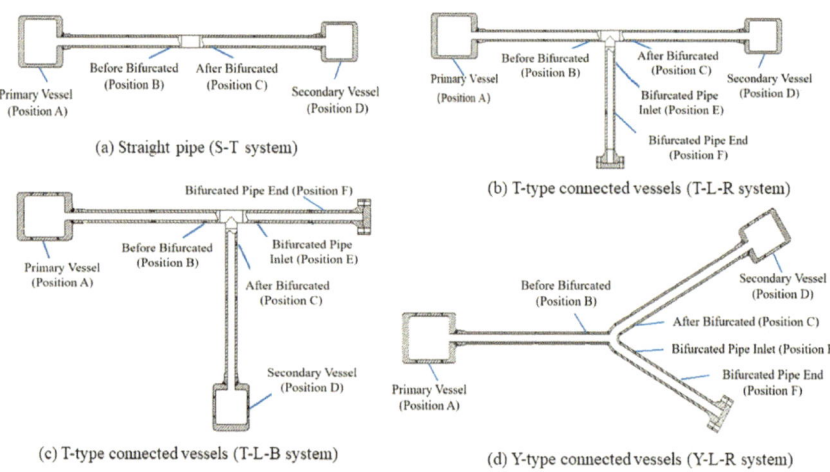

Fig. 3.2 Connection form of vessels and pipes

Table 3.1 Experimental scheme for methane explosion characteristics of no-bifurcated pipe

Influence factor	Experimental parameters					
Bending angle	Primary vessel (V = 22 L), secondary vessel (V = 11 L), cylindrical vessel, pipe length (L = 2 m), central bending of pipe					
	A1	A2	A3	A4	A5	A6
	30°	45°	60°	90°	120°	180°
Bending position	Primary vessel (V = 22 L), secondary vessel (V = 11 L), cylindrical vessel, pipe length (L = 6 m), bending angle ($\theta = 90°$)					
	L1	L2	L3	L4		
	0 m	1 m	3 m	5 m		
Vessel structure	Primary vessel (V = 55 L cylindrical vessel), straight pipe connection (L = 2 m)					
	Secondary vessel	SV1	SV2	SV3	SV4	
		22 L spherical	22 L cylindrical	113 L spherical	113 L cylindrical	

Table 3.2 Effect of the primary vessel structure on P_{max} and $(dP/dt)_{max}$ of the two vessels

Secondary vessel		55 L cylindrical vessel			
Primary vessel		22 L spherical	22 L cylindrical	113 L spherical	113 L cylindrical
P_{max} (MPa)	Primary vessel	0.67	0.63	0.71	0.69
	Secondary vessel	0.76	0.65	1.13	0.86
$(dP/dt)_{max}$ (MPa s^{-1})	Primary vessel	32.5	29.1	27.4	24.9
	Secondary vessel	37.4	31.0	34.6	29.7

The premixed gas of a certain concentration (9.5% vol% of methane) was prepared inside the connected vessels. To ensure good mixing of methane and air, the experimental apparatus was filled with premixed gas by an automatic gas distribution device after achieving a certain vacuum degree (RCS20006). The premixed gas was stood for 30.0 min to ensure uniform mixing. And the absolute pressure of premixed gas before ignition was 0.10 MPa. The overpressure and flame signal data were obtained and analyzed by the program control and data acquisition system. Each experiment of the same condition was repeated 5–6 times to avoid the effects of accidental factors and ensure result accuracy.

3.2 Effect of Connected Vessels Structure on Gas Explosion

3.2.1 Pipe Bending Angle

Straight pipe (S-T system) was adopted to study the influence factor of methane explosion characteristics inside the connected vessels, as shown in Fig. 3.2a. The experimental scheme was shown in Table 3.1. Figure 3.3 shows the overpressure histories at different positions (Position A, B, C and D) inside the 90° bending angle connected vessels with time. The P_{max} before the bending point ($P_{(B)max} = 0.76$ MPa) presented an increasing trend compared with the P_{max} inside the primary vessel ($P_{(A)max} = 0.72$ MPa). It was because that the explosion flame inside the primary vessel after ignition entered the pipe and accelerated along pipe due to the self-acceleration of flame inside the pipe [1]. The overpressure did not change obviously ($P_{(C)max} = 0.76$ MPa) due to the combined action of energy loss caused by the wall blocking (weakening flame propagation capacity) and the flame enhancement caused by the turbulence effect of bending pipe (accelerating flame propagation) [2]. The pressure inside the secondary vessel was increased significantly ($P_{(D)max} = 1.09$ MPa) due to the acceleration of the flame inside the pipe and the enhancement of flame disturbance caused by the sudden increase of propagation space as the flame discharged to secondary vessel. An additional effect which also affects the pressure increase in the secondary vessel was the compression of gases ahead the flame front inside the secondary vessel. The pressure history inside the primary vessel was smooth without obvious oscillation phenomenon. However, the propagations of flame front and pressure wave could be significantly affected due to the action of bending pipe, resulting in an obvious oscillation phenomenon during the pressure history rise at bending position (Position B and C). Besides, the pressure history inside the secondary vessel appeared a periodic oscillation during the descending stage. It was because that the coupling effect between the pressure wave and flame front was enhanced due to the sudden increase of propagation space as the flame discharged to the secondary vessel, resulting in the trend of pressure history periodic oscillation decreasing.

Figure 3.4 shows the effects of pipe bending angles on the P_{max} at different positions (Position A, B, C and D) inside the connected vessels. The P_{max} of four positions presented a trend of rising firstly and then decreasing with the decrease of pipe bending angle. Under 90° bending angle, the P_{max} inside the connected vessels (Position A, B, C and D) reached the maximum. P_{max} inside the secondary vessel was greatest under different bending angles, and the change was also most significantly. Enhancement of flow field disturbance effect resulted in the flame propagation behind the pipe bending point was increased obviously, further resulting in a greater pressure inside the secondary vessel. As $\theta < 90°$, P_{max} of pipe bending point back end (Position C) was obviously smaller than those of the primary vessel (Position A) and bending point front end (Position B). As $\theta > 90°$, the P_{max} presented an opposite phenomenon. It was because that a high-pressure zone appeared in the bending point back end as $\theta > 90°$, resulting in the greater pressure in this zone. As the bending angle decreased,

Fig. 3.3 Overpressure
histories at different
positions inside the 90°
bending angle connected
vessels

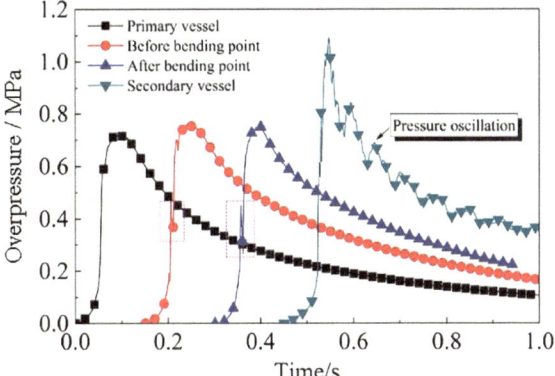

Fig. 3.4 Effect of pipe
bending angle on the P_{max} at
different positions inside the
connected vessels

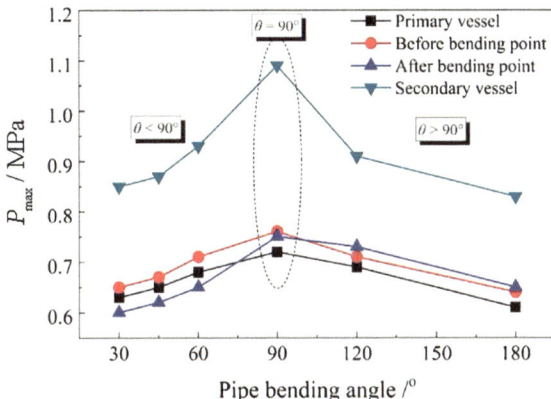

the high-pressure zone gradually moved toward the bending point and crossed the
bending point to its front end. Therefore, the P_{max} of the bending point back end was
lower than that of front end as $\theta < 90°$.

The voltage signal was generated as the explosion flame passed through the flame
detector. Figure 3.5 shows the voltage signal histories at different positions inside
the connected vessels ($\theta = 90°$). The voltage signal rise moments of four positions
(the primary vessel, bending point front end, bending point back end and secondary
vessel) were t_1, t_2, t_3 and t_4, respectively ($t_1 < t_2 < t_3 < t_4$). The time that the flame
propagated from the primary vessel to bending point front end ($\Delta t_1 = 13.0$ ms) was
significantly greater than the time that the flame propagated from the bending point
back end to secondary vessel ($\Delta t_3 = 5.7$ ms), and the time for the flame passed through
the bending point was 4.2 ms (Δt_2). It indicated that the turbulence disturbances of
the flame and unburned gas were enhanced as the flame passed through the bending
point, further resulting in the increase of flame propagation velocity [3].

The v_{ave} under different pipe sections can be obtained through the distance of
adjacent flame detector (Positions A, B, C and D) and the time that flame propagated

Fig. 3.5 Voltage signal
histories at different
positions inside the
connected vessels ($\theta = 90°$)

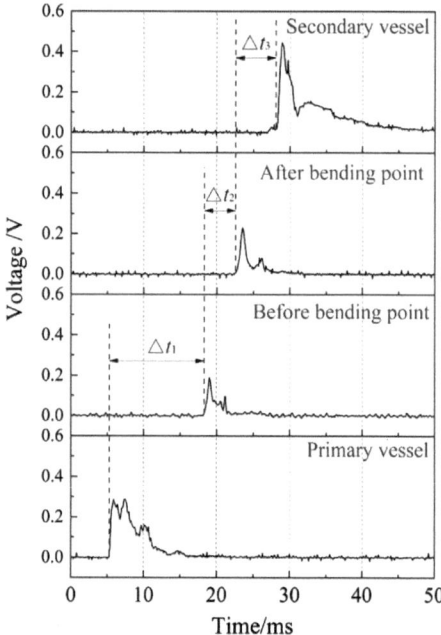

at the two adjacent points. Figure 3.6 shows the various of v_{ave} inside the different pipe sections under six bending angles ($\theta = 30, 45, 60, 90, 120$ and $180°$). The bending angle had a small effect on the v_{ave} between the primary vessel and bending point. It was because that the blocking and reverse propagating of flame and pressure wave by the pipe bending wall surface were enhanced as the decrease of bending angle, resulting in a slight decrease of v_{ave} [4]. And the v_{ave} presented a similar various trend under the bending point. The flame turbulence intensity presented the increase of different degrees due to the effect of bending angle, which resulted in the flame propagation velocity difference. Between the bending point and secondary vessel, the v_{ave} presented a trend of rising firstly and then decreasing with the decrease of bending angle. And the v_{ave} was the greatest under 90° bending angle. It indicated that the combined effects of turbulent disturbance of the pipe on the flow field and the energy absorption of wall surface on the flame resulted in the most significant increase in flame propagation velocity at this bending angle.

3.2.2 Pipe Bending Position

Figure 3.7 shows the effect of pipe bending position (distance from the end of primary vessel (L) = 0, 1, 3 and 5 m) on P_{max} inside the primary and secondary vessels. It can be seen that the P_{max} inside the secondary vessel was significantly greater than that of primary vessel. P_{max} inside the primary vessel showed a slight increase trend with

Fig. 3.6 Effect of pipe bending angle on the v_{ave} inside the connected vessels

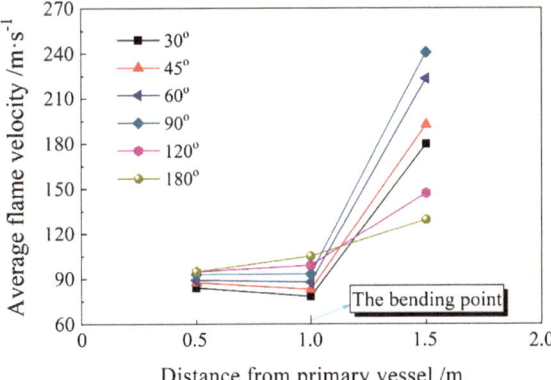

the increase of pipe bending position. However, P_{max} inside the secondary vessel presented a decreasing trend, and it was obviously higher than without bending point. Meanwhile, the overpressure rising rate could be obtained by the derivation of pressure history with time [5]. The $(dP/dt)_{max}$ presented a similar variation trend. As the bending point existed, $(dP/dt)_{max}$ inside the primary vessel was increased by 13% and 6% and $(dP/dt)_{max}$ inside the secondary vessel was decreased by 18% and 23% respectively with the increase of bending position. Nevertheless, the $(dP/dt)_{max}$ inside the secondary vessel was greater compared with the primary vessel. And the both were larger than without bending point $((dP/dt)_{max}$ (primary vessel) $= 38.6$ MPa s^{-1}, $(dP/dt)_{max}$ (secondary vessel) $= 139.0$ MPa s^{-1}). It indicated that the pipe bending could significantly affect the explosion flow field inside the connected vessels and the flame propagation inside the pipe. And the disturbed flame was accelerated significantly inside the pipe, resulting in a higher pressure inside the secondary vessel with increase of the distance between the bending point and secondary vessel.

Fig. 3.7 Effect of pipe bending position on the P_{max} inside the connected vessels

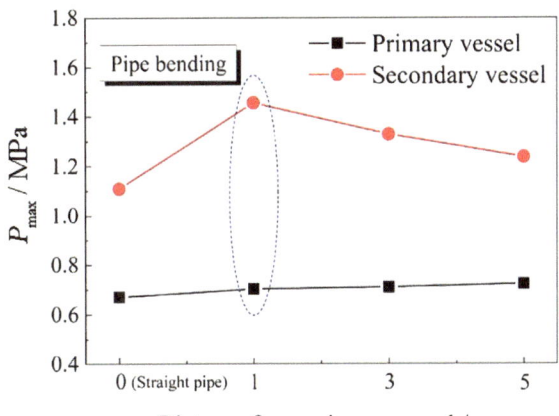

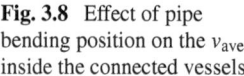

Fig. 3.8 Effect of pipe bending position on the v_{ave} inside the connected vessels

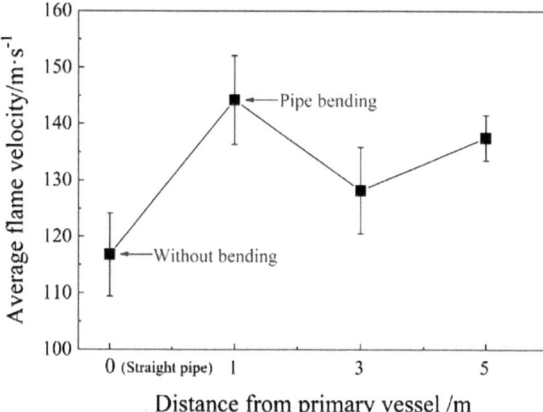

Figure 3.8 presents the effect of pipe bending position on the v_{ave} inside the connected vessels. It can be seen that the v_{ave} was increased significantly as the pipe bent. Flame propagation velocity reached maximum inside the connected vessels as the bending position was far away from the end of secondary vessel ($L = 1$ m). It indicated that the length of turbulent flame accelerating propagation was increased inside the pipe as the increase of distance between the bending position and secondary vessel, resulting in a higher flame propagation rate. With the reduction of distance ($L = 3$ and 5 m), the length of turbulent flame accelerating propagation was decreased inside the pipe. And it further resulted in the reduction of v_{ave}. As the pipe bending position moved back, the 5 m bending position showed a greater v_{ave} compared with the 3 m bending position due to a combination of flame self-acceleration inside the pipe and turbulent disturbance caused by the bending pipe.

3.2.3 Vessel Shape and Volume

The effect of vessel structure on the overpressure inside the connected vessels was studied by changing the primary and secondary vessel structures (volume and shape), as shown in Fig. 3.2a. The experimental scheme was shown in Table 3.1. Figure 3.9a, b shows the effect of secondary vessel structure on P_{max} and $(dP/dt)_{max}$ inside the primary and secondary vessels.

As can be seen from Fig. 3.9a, the shape of the secondary vessel had a small effect on the P_{max} inside the primary vessel, but it had a larger effect on the P_{max} inside the secondary vessel under the same secondary vessel volume. Meanwhile, the P_{max} inside the spherical secondary vessel was larger compared with the cylindrical secondary vessel. Under the same secondary vessel shape, the P_{max} inside the large volume secondary vessel corresponding to the primary and secondary vessels was larger than that of the small volume secondary vessel. It indicated that the volume of

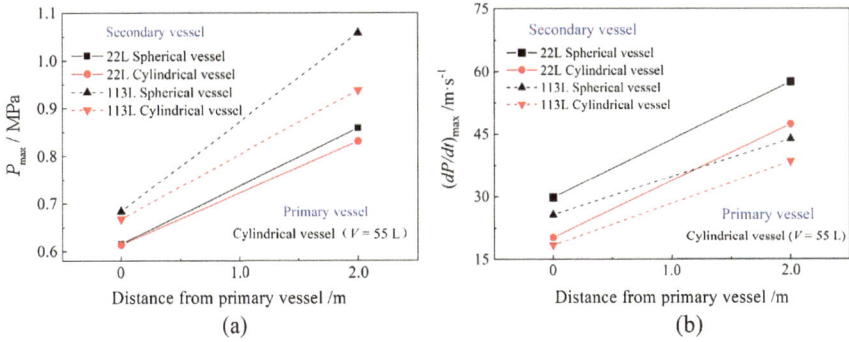

Fig. 3.9 a Effect of the secondary vessel structure on the P_{max} inside the primary and secondary vessels. **b** Effect of the secondary vessel structure on the $(dP/dt)_{max}$ inside the primary and secondary vessels

the secondary vessel had a greater effect on the explosion intensity inside the primary and secondary vessels. The effect extent of secondary vessel shape was related to its volume. The effect of larger volume secondary vessel was more significant, and the effect of spherical vessel was greater than that of cylindrical vessel. As can be seen from Fig. 3.9b, the spherical secondary vessel corresponding to the $(dP/dt)_{max}$ inside the primary and secondary vessels was larger under the same volume secondary vessel compared with the cylindrical secondary vessel. Under the same shape secondary vessel, the small volume secondary vessel (V = 22 L) corresponding to the $(dP/dt)_{max}$ inside the primary and secondary vessels was larger compared with the large volume secondary vessel (V = 113 L). It indicated that the small volume secondary vessel had a large effect on the pressure rise inside the primary and secondary vessels, and that the effect of spherical secondary vessel was more significant.

Table 3.2 shows the effect of primary vessel structure on the P_{max} and $(dP/dt)_{max}$ inside the primary and secondary vessels. As can be seen from Table 3.2, the P_{max} of the spherical primary vessel corresponding to the primary and secondary vessels was larger under the same volume primary vessel compared with the cylindrical vessel. And the change of P_{max} inside the secondary vessel was more significant. Under the same shape primary vessel, the large volume primary vessel (V = 113 L) corresponding to the P_{max} was larger compared with the small volume primary vessel (V = 22 L). And the change of P_{max} inside the secondary vessel was also more significant. It indicated that the structure of primary vessel had a greater impact on the pressure inside the secondary vessel, and it was shown that the large volume spherical primary vessel had the greatest effect on the explosion intensity inside the secondary vessel. Meanwhile, the small volume primary vessel corresponding to the $(dP/dt)_{max}$ inside the primary and secondary vessels was larger compared with the large volume primary vessel under the same shape primary vessel. Under the same volume primary vessel, the effect of spherical primary vessel on the pressure rising rate was greater. And it was shown that the spherical primary vessel corresponding to the $(dP/dt)_{max}$ was larger. Above results showed that the spherical primary vessel had a greater effect on the pressure rise inside the primary and secondary vessels, and the effect of small volume vessel was more significant.

3.3 Effect of Bifurcated Structure on Gas Explosion

3.3.1 T-L-R Connected Vessels

A T-L-R type connected vessels was adopted to study the effect of bifurcated pipe on the methane explosion inside the connected vessels and the experimental apparatus was shown in Fig. 3.2b. The experimental scheme was shown in Table 3.3. Figure 3.10 shows the overpressure histories at different positions inside the T-L-R connected vessels with time. The pressure histories of primary vessel (Position A), the bifurcated front (Position B) and back (Position C) were similar. However, the pressure was increased slightly due to the acceleration effect of pipe for the explosion flame (from 0.62 to 0.65 MPa). Compared with the non-bifurcated pipe (in Fig. 3.3), the pressure inside the secondary vessel was significantly reduced from 1.09 to 0.90 MPa. It was because that most of the flame and pressure wave propagated to the secondary vessel along the main pipe and part of them propagated to the bifurcated pipe, resulting in the reduction of the energy propagated to secondary vessel. Meanwhile, the pressure histories at the bifurcated point (Position E) and end (Position F) was almost the similar. But it was significantly lower than the pressures inside the main pipe and secondary vessel. This also indicated that the flame propagation energy entered bifurcated pipe was lower than that of main pipe. Due to the deformation and superposition of flame and pressure wave, an obvious pressure history oscillation appeared at the bifurcated point (Position E). Besides, the pressure history inside the bifurcated pipe end (Position F) appeared a rebound phenomenon (in Fig. 3.10). It was because that the explosion flame and pressure wave occurred accumulation and back-propagation due to the blocking of flange end [6].

Figure 3.11 shows the voltage signal histories at different positions inside the T-L-R connected vessels with time. The signal was detected by the flame detector inside the primary vessel at 5.2 ms after ignition. At 16.7 ms, the flame propagated to bifurcated pipe front, which was the initial stage of flame propagation with a slow propagation velocity. After that, the flame rapidly propagated to the Position C due to the disturbance effect of bifurcated pipe. And the flame inside the bifurcated pipe

Table 3.3 Experimental scheme for gas explosion characteristics of bifurcated pipe

Experimental apparatus	Primary vessel ($V = 22$ L), secondary vessel ($V = 11$ L), cylindrical vessel, pipe length ($L = 2$ m), central bending of pipe, bifurcated pipe length ($L = 1$ m)			
Connection form	S1	S2	S3	S4
	S-T system	T-L-R system	T-L-B system	Y-L-R system

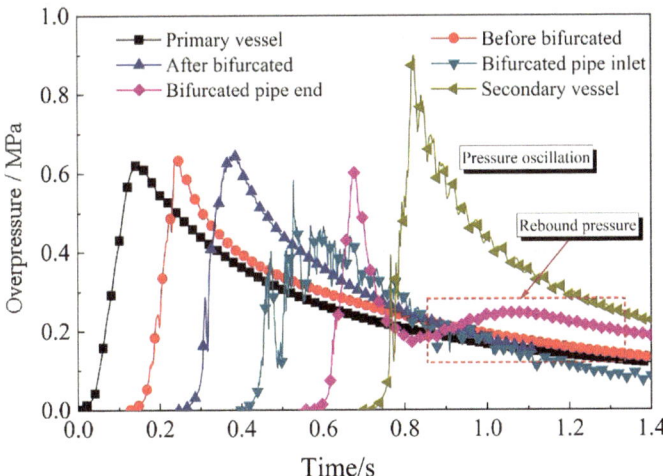

Fig. 3.10 Overpressure at different positions inside the T-L-R connected vessels with time

also propagated to the Position E at this moment. The two points corresponding to the rise moment of signal history were 19.9 ms and 21.7 ms respectively. The times for the flame reached the bifurcated pipe end and secondary vessel were almost the same (t = 27.0 and 27.6 ms). Besides, the voltage signal history of bifurcated pipe end appeared two obvious rise processes (in Fig. 3.11). It indicated that the flame occurred back-propagation due to the blocking of flange end. This further explained that the pressure rebounded at the bifurcated pipe end (Position F) resulted from the back propagation of flame (in Fig. 3.10).

3.3.2 T-Type Connected Vessels

The T-type connected vessels (S-T, T-L-R and T-L-B systems) was adopted to study the effect of bifurcated pipe connection form on methane explosion, as shown in Fig. 3.2a–c. The experimental scheme was shown in Table 3.3. Figure 3.12a shows the effect of bifurcated pipe connection form on the P_{max} at different positions (Position A–F). The three connection forms had a small impact on the P_{max} of Position A, B and C inside the connected vessels. However, it had a great impact on the P_{max} of Position D, E and F inside the bifurcated pipe and secondary vessel. For the P_{max} inside the secondary vessel, T-L-R connection form showed the greatest P_{max}, followed by the S-T connection form. And the smallest P_{max} appeared in T-L-B connection form. Meanwhile, the P_{max} inside the bifurcated pipe (Position E and F) of T-L-B connection form was larger (i.e. the horizontal bifurcated pipe) compared with the T-L-R connection form. It indicated that the connection form of connected vessels had a great impact on the pressures inside the secondary vessel and bifurcated

Fig. 3.11 Voltage signal histories at different positions inside the T-L-R connected vessels

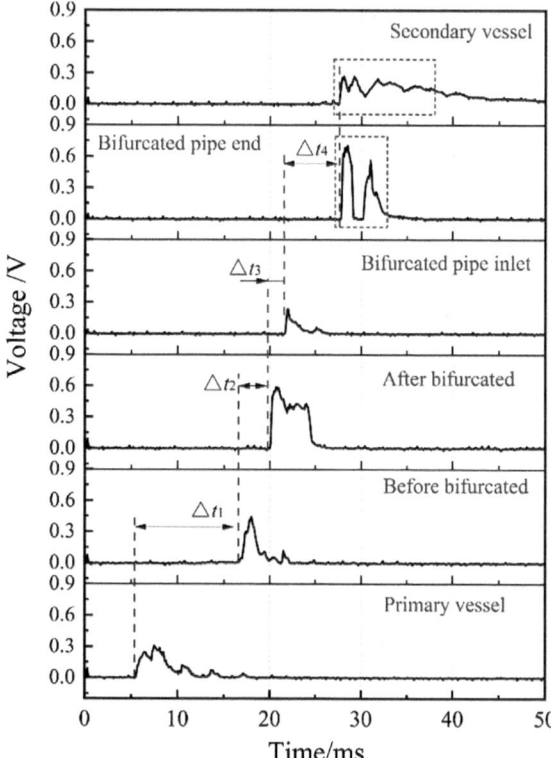

pipe, and it had a small impact on the pressure inside the main pipe of bifurcated point front. The above phenomenon was due to the shunt effect of bifurcated pipe. Part of the flame and pressure wave would change the propagation direction and propagate along the bending pipe, while most of the flame and pressure wave would keep the initial direction and continue to propagate along the horizontal pipe. This resulted in a higher overpressure inside the horizontal pipe and a lower overpressure inside the bending pipe. Meanwhile, it was shown that P_{max} inside the secondary vessel of T-L-R connection form was significantly greater than that of S-T connection form due to the combined effects of turbulence enhancing and shunt weakening inside the bifurcated pipe on the explosion flow field [7].

The effect of bifurcated pipe on explosion intensity could be directly reflected by the overpressure rising rate. Figure 3.12b shows the effect of T-type connection form on the $(dP/dt)_{max}$ at different positions inside the connected vessels (Position A–F). The greatest $(dP/dt)_{max}$ appeared in the four positions (A–D) of S-T connection form, followed by the T-L-R connection form. And the smallest $(dP/dt)_{max}$ appeared in the T-L-B connection form. It indicated that the pressure rising rate inside the connected vessels was greatest without bifurcated pipe. This was because that the shunt of part of flame and pressure wave resulted in the weakening of flame and pressure wave energy during the propagation due to the presence of bifurcated pipe. However,

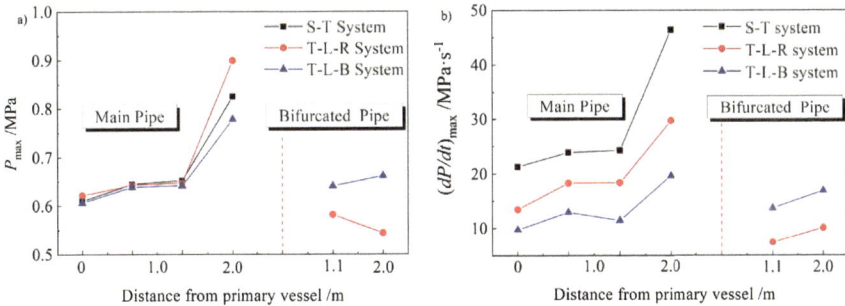

Fig. 3.12 a Effect of bifurcated pipe connection form on the P_{max} inside the connected vessels (Position A, B, C, D, E and F). **b** Effect of bifurcated pipe connection form on the $(dP/dt)_{max}$ inside the connected vessels (Position A, B, C, D, E and F)

most of the flame and pressure wave would still maintain the initial direction and continue to propagate along the horizontal pipe although shunt effect occurred. It made that the $(dP/dt)_{max}$ at the four Positions (A–D) of T-L-R connection form was significantly larger than that of T-L-B connection form. It was also shown that the $(dP/dt)_{max}$ of bifurcated pipe (Position E and F) in the T-L-B connection form (i.e. the horizontal bifurcated pipe) was significantly greater than that of T-L-R connection form corresponding two positions.

Figure 3.13 shows a comparison of the v_{ave} of each section inside the connected vessels. The v_{ave} of A–B and C–D sections inside the T-L-R connected vessels was significantly larger than that of the corresponding section inside the T-L-B and S-T connected vessels. The v_{ave} of A–B section inside the T-L-B and S-T connected vessels was almost the same. However, the v_{ave} of C–D section inside the S-T connected vessels was slightly increased. It was because that the reverse restraint effect of S-T connection form secondary vessel on the flame and pressure wave inside the C–D pipe was smaller, so that the flame could be rapidly discharged to a relatively large space after accelerating propagation inside the pipe. The energy propagated to secondary vessel was reduced due to the shunt effect of bifurcated pipe although the disturbance of bifurcated pipe on the flow field could enhance the flame propagation velocity, thus showing a slight reduction in flame propagation rate under the C–D section of T-L-B connection form. As the bifurcated pipe existed, most of the flame and pressure wave propagation would remain along the initial direction (i.e. horizontal pipe propagation). Therefore, the v_{ave} of the E–F section inside the T-L-B bifurcated pipe was significantly greater than that of corresponding to the T-L-R bifurcated pipe.

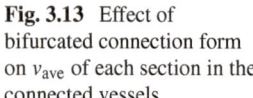

Fig. 3.13 Effect of bifurcated connection form on v_{ave} of each section in the connected vessels

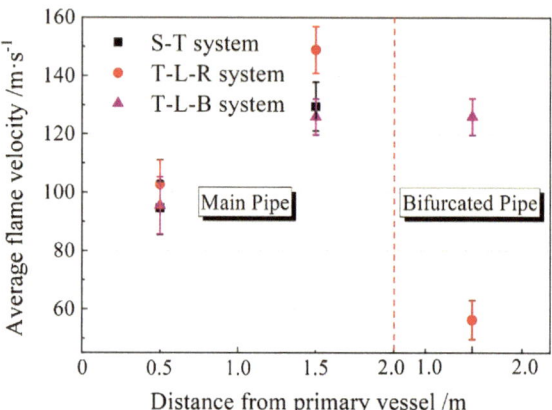

3.3.3 Comparison of T and Y Connected Vessels

T and Y connected vessels (S-T, T-L-R and Y-L-R system) were adopted to study the effect of bifurcated pipe angle on the methane explosion, as shown in Fig. 3.2a, b and d. And the experimental scheme was shown in Table 3.3. Figure 3.14a shows the effect of bifurcated pipe angle on the P_{max} inside the connected vessels. The P_{max} of T-L-R connection form was greatest, and the P_{max} of S-T connection form was smallest inside the secondary vessel under different connection forms. It indicated that the increase of bifurcated pipe energy shunt (the increase of bending angle) resulted in an obvious reduction in the energy propagated along the main pipe due to the presence of Y-L-R connected vessels main pipe bending angle compared with the T-L-R connected vessels. Effects of different connection forms on the P_{max} inside the primary vessel (Position A) and the main pipe (Position B and C) were not significant. However, the P_{max} inside the connected pipe (Position E and F) was greater under the Y-type (Y-L-R system) connection form compared with T-type (T-L-R system) connection form. This indicated that the energy shunt of Y-type (Y-L-R system) connected vessels bifurcated pipe was increased as it tended to be parallel to the main pipe, which showing a larger P_{max} [8]. This also indicated that the presence of bifurcated pipe could affect the energy shunt of explosion flow field and the shunt extent was related to the pipe bending angle, further affecting the overpressure inside the pipe.

Figure 3.14b shows the effect of bifurcated pipe angle on the $(dP/dt)_{max}$ inside the connected vessels. The $(dP/dt)_{max}$ was the greatest at the main pipe four positions (Position A, B, C and D) inside the S-T connected vessels, followed by the T-L-R vessel. And the smallest value appeared in the main pipe inside the Y-L-R vessel. It indicated that the energy of explosion flow field inside the main pipe was reduced due to the presence of the bifurcated pipe, further resulting in an obvious reduce of pressure rising rate. Meanwhile, the presence of main pipe bending angle and its increase would further reduce the energy of flow field inside the main pipe,

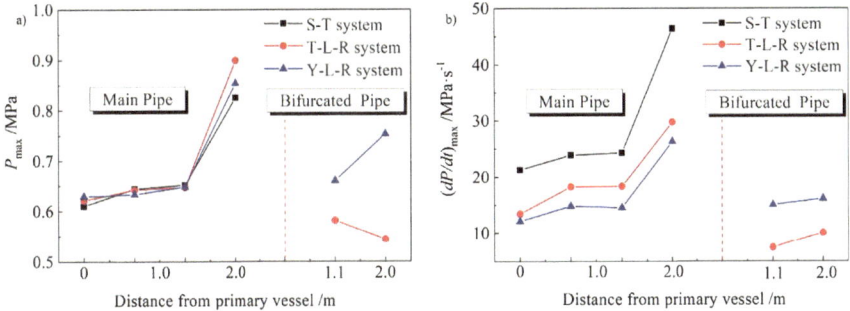

Fig. 3.14 a Effect of bifurcated pipe angle on the P_{max} inside the connected vessels. **b** Effect of connected pipe angle on the $(dP/dt)_{max}$ inside the connected vessels

resulting a smallest $(dP/dt)_{max}$ appearance inside the Y-L-R connected vessels main pipe. For the T and Y bifurcated pipes, the $(dP/dt)_{max}$ inside the Y-L-R connected vessels bifurcated pipe (Position E and F) was significantly greater than that of T-L-R connected vessels, which also indicated that energy shunt was an important effect factor for the pressure rising rate inside the pipe.

Figure 3.15 shows the effect of bifurcated pipe angle on the v_{ave} of each section inside the connected vessels. The v_{ave} inside the A–B and C–D sections of T-type (T-L-R system) connected vessels main pipe were greater than those of the corresponding positions inside the Y-type (Y-L-R system) connected vessels main pipe, and the smallest v_{ave} appeared in the S-T connected vessels main pipe. However, the v_{ave} inside the Y-type connected vessels bifurcated pipe was greater compared with the T-type connected vessels. This indicated that the turbulence effect of different extent resulted from the bifurcated pipe structure was an important influence factor for the flame propagation rate. Besides, the bifurcated pipe structure could also lead to the obvious change of interaction among compression wave, sparse wave and reflection wave at the pipe bending position.

The explosion flow field at the pipe bending position could be affected by the interaction among compression wave, sparse wave and reflection wave [9]. A local high-pressure zone was formed at the concave wall surface of bending position due to the superposition of compression and reflection waves, which blocking the flame propagation [10]. Meanwhile, the high-pressure zone moved at the pipe bending position with the change of bending angle, which further affecting the flow field. At the convex wall surface of pipe bending position, the flow velocity of premixed gas was accelerated due to the stretching effect of sparse wave [11]. Hence, the flame front occurred deformation and wrinkle, resulting in the increases of flame surface area and methane combustion rate under the combined effect of multiple wave systems. And the flame propagation velocity and overpressure were gradually enhanced with the increase of flame front turbulence intensity. The pipe bending position could affect the moment of flame turbulent acceleration propagation, further affecting the explosion intensity. Meanwhile, the flame and pressure wave were blocked by the pipe bending wall surface and propagated reversely, which also significantly affect

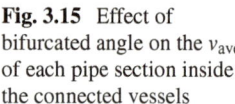
Fig. 3.15 Effect of bifurcated angle on the v_{ave} of each pipe section inside the connected vessels

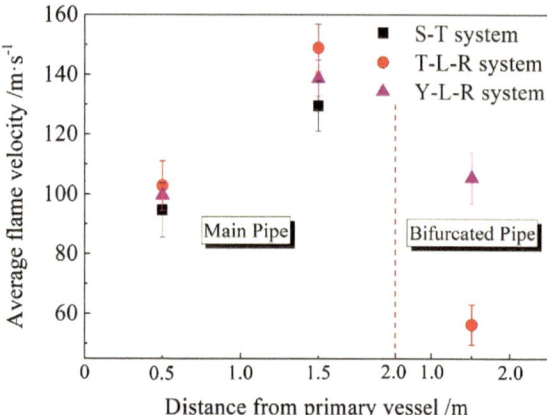

the explosion intensity inside the pipe. Besides, part of free radicals produced from the reaction process would be terminated due the wall effect after impacting with the wall surface according to the theory of branched chain reaction (H·, OH·, CH$_3$O· and CH$_3$·) [12, 13]. The amount of free radical destruction could be changed by influencing the area and frequency of flame collision with the wall surface due to the changes of pipe bending angle, position and structure, as shown in Fig. 3.16. Hence, different explosion intensity was presented inside the connected vessels under the combination of above effects.

The existence of bifurcated pipe could result in the energy shunt of explosion flow field inside the main pipe. And the shunt extent was related to the bifurcated pipe angle and connection form, which further affected the explosion intensity inside the main and bifurcated pipes. Part of flame and pressure wave could change the propagation direction and propagate along the bending pipe, while most of them will maintain the initial direction and continue to propagate along the horizontal pipe as the bifurcated pipe existed [14]. It made a higher pressure in the horizontal pipe than that in the bending pipe. The energy in flame propagation along the main pipe

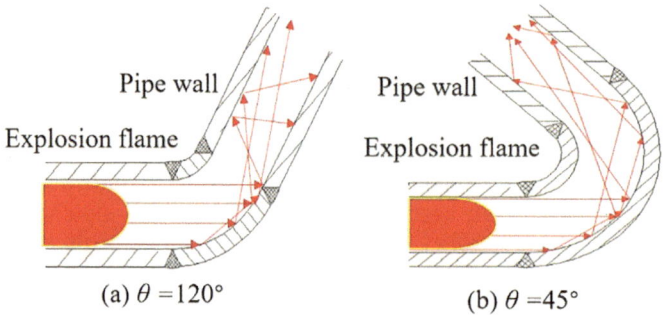

Fig. 3.16 Schematic diagram of free radical propagation collision at pipe bending position

was weakened due to the shunt effect of bifurcated pipe. However, the presence of bifurcated pipe would also affect the flow field inside the pipe and generate a greater turbulent disturbance effect [7, 15], further resulting in the significant increases of flame propagation velocity and overpressure. Hence, the effect of bifurcated pipe on the explosion intensity inside the connected vessels depended on the combined effect of the turbulent disturbance enhancing and the shunt weakening on the explosion flow field.

References

1. Daubech, C. L. G. (2019). Further insight into the gas flame acceleration mechanisms in pipes. Part I: Experimental work. *Journal of Loss Prevention in the Process Industries, 62*, 103930.
2. Guoxun, J., Guo, S., Zhiwei, J., et al. (2008). Experimental study on propagation regulation of gas explosive shock wave at turning point of tunnel. In *The 2008 International Symposium on Safety Science and Technology*.
3. Kundu, S. K., Zanganeh, J., Eschebach, D., et al. (2017). Explosion characteristics of methane–air mixtures in a spherical vessel connected with a duct. *Process Safety and Environmental Protection, 111*, 85–93.
4. Phylaktou, H., & Andrews, G. E. (1993). Gas explosions in linked vessels. *Journal of Loss Prevention in the Process Industries, 6*(1), 15–19.
5. Cao, X., Bi, M., Ren, J., et al. (2019). Experimental research on explosion suppression affected by ultrafine water mist containing different additives. *Journal of Hazardous Materials, 368*, 613–620.
6. Cao, X., Ren, J., Bi, M., et al. (2017). Experimental research on the characteristics of methane/air explosion affected by ultrafine water mist. *Journal of Hazardous Materials, 324*, 489–497.
7. Lin, B., Guo, C., Sun, Y., et al. (2016). Effect of bifurcation on premixed methane-air explosion overpressure in pipes. *Journal of Loss Prevention in the Process Industries, 43*, 464–470.
8. Jing, G., Guo, S., & Wu, Y. (2020). Investigation on methane-air explosion overpressure in bifurcated tubes according to methane concentrations and bifurcation angles. *Energy Sources, Part A: Recovery, Utilization, and Environmental Effects*, 1–12.
9. Zhou, N., Ni, P. F., Li, X., et al. (2021). Study on influence of pipe structure on explosion characteristics of hydrogen-air premixed gas. *Journal of Safety Science and Technology, 17*, 65–71.
10. Zhou, N., Wang, W. X., Zhang, G. W., et al. (2018). Numerical simulation study on the combustion rule of bending structure in pipes. *Combustion Science and Technology, 190*(9), 1500–1514.
11. Xiao, H., He, X., Wang, Q., et al. (2013). Experimental and numerical study of premixed flame propagation in a closed duct with a 90° curved section. *International Journal of Heat and Mass Transfer, 66*, 818–822.
12. Dreizler, A., & Böhm, B. (2015). Advanced laser diagnostics for an improved understanding of premixed flame-wall interactions. *Proceedings of the Combustion Institute, 35*(1), 37–64.
13. Salimath, P. S., Ertesvåg, I. S., & Gruber, A. (2020). Computational analysis of premixed methane-air flame interacting with a solid wall or a hydrogen porous wall. *Fuel, 272*, 117658.
14. Li, G., Du, Y., Qi, S., et al. (2016). Explosions of gasoline-air mixtures in a closed pipe containing a T-shaped branch structure. *Journal of Loss Prevention in the Process Industries, 43*, 529–536.
15. Razus, D., Oancea, D., Chirila, F., et al. (2003). Transmission of an explosion between linked vessels. *Fire Safety Journal, 38*(2), 147–163.

Chapter 4
Scale Effect of Gas Explosion in the Confined Space

4.1 Experimental Apparatus and Methods

4.1.1 Experimental Apparatus

Based on the linked gas explosion system, the scale effect on premixed methane–air explosion in linked device was studied. The physical diagram and schematic of experimental apparatus were shown in Figs. 4.1 and 4.2.

The vessels and connected pipes were consist of 16 Mn III vessel steel, with a design pressure of 20 MPa. The flange opening can be sealed with a blind plate (material Q345R). The volume of cylindrical vessels was 11, 22, 55 and 113 L with an aspect ratio of 1:1. The length of each pipe was 2 m, and the inner diameters were 20, 59, 108 and 133 mm. Six kinds of gradient pipes with different specifications were used in the experiments to change the inner diameter of the pipe. A 2X-8GA vacuum pump and an RCS2000-B gas dispenser were used to configure the gas mixture in the experimental setup. An adjustable high-energy igniter, type KTD-A, was used for ignition, with an ignition energy range of 5–20 J. High-frequency dynamic pressure sensors, type HM90H2-2-V2-F10-W2, were mounted on the walls of the vessel and pipes to collect the overpressure. A DEWE SIRIUS type data acquisition instrument was used for data acquisition.

4.1.2 Experimental Methods

Firstly, connect the vessel and the pipe through flanges and fasteners, install sensors, ignition rods, inlet and outlet valves, then check the air tightness of the experimental system and the status of each part. Turn on the vacuum pump and reduce the pressure inside the experimental apparatus to − 0.09 MPa. Methane of 10% concentration was fed into the experimental setup through the gas mixer. When the pressure reaches

Z. Wang and X. Cao, *Gas Explosion and Its Protection Technology in Process Industries*, https://doi.org/10.1007/978-981-96-3121-6_4

Fig. 4.1 Physical diagram

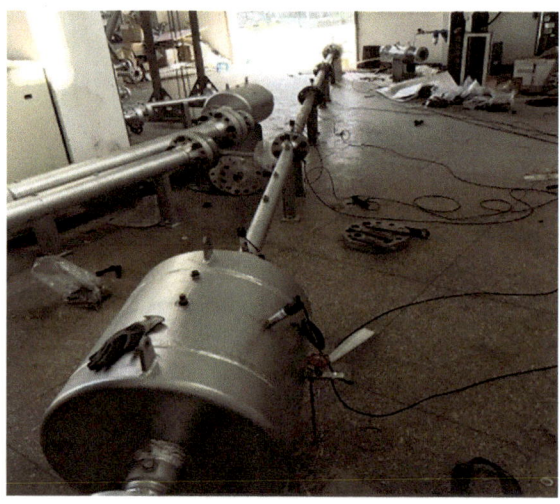

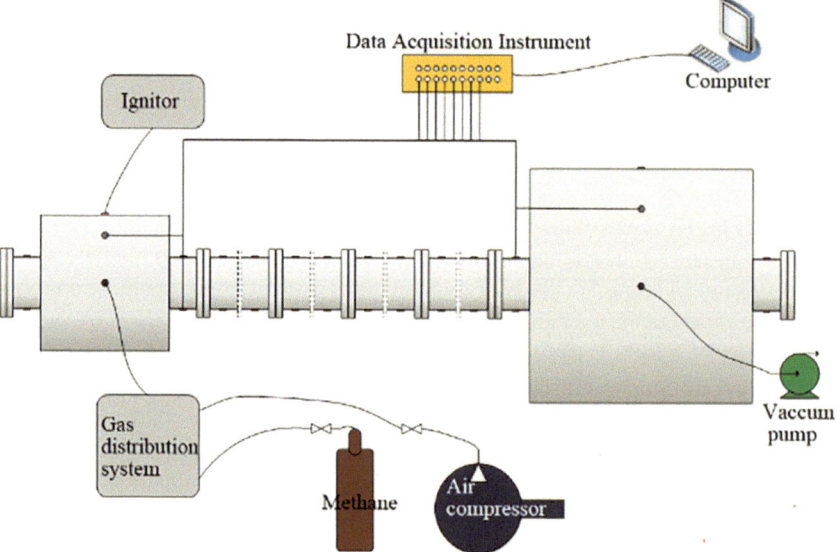

Fig. 4.2 Schematic diagram

0 MPa, turn off the gas mixer and let it sit for 5 min. Then ignite the gas and collect the overpressure. Opening the inlet and outlet valves of the device, using the air compressor to purge the experimental device, and then proceeding to the next set of experiments.

4.1.3 Experimental Conditions

The experiment was carried out at normal temperature (25 °C) and pressure (0.1 MPa). Methane–air mixture (10%) was adopted during the experiment. Ignition position was in the center of the vessel, and the ignition energy was 5 J. The specific experimental program was shown in Tables 4.1, 4.2 and 4.3. This experiment mainly studies the effects of changes in duct length, duct inner diameter, and volume ratio of primary and secondary vessel (V_1/V_2) on the explosion characteristics of premixed methane–air in the linked vessel. Each group of experiments was repeated three times, and the average value was taken to ensure the accuracy of the experimental data. The schematic diagrams of the experimental device structure were shown in Figs. 4.3 and 4.4.

4.2 Scale Effect of Single Vessel Structure

In this section, four cylindrical vessels of different volumes (11, 22, 55 and 113 L) were selected, and a central ignition method was adopted. The pressure transducers were installed at the center of the side wall of the vessel. Figure 4.5 shows the change in the overpressure of premixed methane–air in cylindrical vessels with different volume, and the maximum overpressure (P_{max}) and maximum rate of pressure rise ($(dP/dt)_{max}$) were shown in Table 4.4. It can be observed in Fig. 4.5 that the overpressure in four vessels of different volumes increased sharply and rapidly decreases upon reaching P_{max}. According to Table 4.4, with the increase of vessel volume, the P_{max} inside the vessel increased from 0.63 to 0.66 MPa, while the $(dP/dt)_{max}$ decreased from 12.64 to 5.96 MPa s^{-1}.

Table 4.1 Single vessel structure

No.	Selection of vessel	Position of sensor
K01	11 L vessel	Central wall of the vessel
K02	22 L vessel	Central wall of the vessel
K03	55 L vessel	Central wall of the vessel
K04	113 L vessel	Central wall of the vessel
K05	L = 2 m, D = 59 mm pipe	One end of pipe
K06	L = 4 m, D = 59 mm pipe	Pipe end
K07	L = 6 m, D = 59 mm pipe	Pipe end
K08	L = 8 m, D = 59 mm pipe	Pipe end
K09	L = 2 m, D = 20 mm pipe	Pipe end
K10	L = 2 m, D = 108 mm pipe	Pipe end
K11	L = 2 m, D = 133 mm pipe	Pipe end

Table 4.2 Vessel–pipe structure

No.	Volume of vessel (L)	Pipe length (m)	Pipe inner diameter (mm)	Position of sensor
K12	11	2	20	Central wall of the vessel + pipe end
K13	11	2	59	Central wall of the vessel + pipe end
K14	11	2	108	Central wall of the vessel + pipe end
K15	11	2	133	Central wall of the vessel + pipe end
K16	11	4	59	Central wall of the vessel + pipe end
K17	11	6	59	Central wall of the vessel + pipe end
K18	11	8	59	Central wall of the vessel + pipe end
K19	22	2	20	Central wall of the vessel + pipe end
K20	22	2	59	Central wall of the vessel + pipe end
K21	22	2	108	Central wall of the vessel + pipe end
K22	22	2	133	Central wall of the vessel + pipe end
K23	22	4	59	Central wall of the vessel + pipe end
K24	22	6	59	Central wall of the vessel + pipe end
K25	22	8	59	Central wall of the vessel + pipe end
K26	55	2	20	Central wall of the vessel + pipe end
K27	55	2	59	Central wall of the vessel + pipe end
K28	55	2	108	Central wall of the vessel + pipe end
K29	55	2	133	Central wall of the vessel + pipe end
K30	55	4	59	Central wall of the vessel + pipe end
K31	55	6	59	Central wall of the vessel + pipe end
K32	55	8	59	Central wall of the vessel + pipe end

(continued)

Table 4.2 (continued)

No.	Volume of vessel (L)	Pipe length (m)	Pipe inner diameter (mm)	Position of sensor
K33	113	2	20	Central wall of the vessel + pipe end
K34	113	2	59	Central wall of the vessel + pipe end
K35	113	2	108	Central wall of the vessel + pipe end
K36	113	2	133	Central wall of the vessel + pipe end
K37	113	4	59	Central wall of the vessel + pipe end
K38	113	6	59	Central wall of the vessel + pipe end
K39	113	8	59	Central wall of the vessel + pipe end

Based on the experimental results, the following quantitative prediction models were obtained:

$$P_{max} = 0.66695 - 0.05175 \times 0.96791^V, R^2 = 0.99863 \qquad (4.1)$$

$$(dP/dt)_{max} = 5.74098 + 9.64086 \times 0.96902^V, R^2 = 0.99393 \qquad (4.2)$$

where V (L) was vessel volume. The confidence coefficient R^2 were both greater than 0.9. Therefore, it can effectively predict the P_{max} and $(dP/dt)_{max}$ of methane–air gas explosion in vessels of different volumes.

According to the data in Table 4.4 and the cube root law, the explosion indexes K_g were 2.81 MPa m s^{-1}, 2.91 MPa m s^{-1}, 2.88 MPa m s^{-1} and 2.88 MPa m s^{-1}, respectively. Due to certain experimental errors, it can be assumed that the overpressure of premixed methane–air in this cylindrical vessel conforms to the "root of the cube" law. Here, the average value of 2.87 MPa m s^{-1} was taken as its explosion index.

Table 4.3 Vessel–pipe–vessel structure

No.	V_1/V_2	V_1 (L)	V_2 (L)	Pipe length (m)	Pipe inner diameter (mm)	Position of sensor
K40	0.1	11	113	2	59	Central wall of the vessel
K41	0.2	22	113	2	59	Central wall of the vessel
K42	0.4	22	55	2	59	Central wall of the vessel
K43	0.5	11	22	2	20	Central wall of the vessel
K44	0.5	11	22	2	59	Central wall of the vessel
K45	0.5	11	22	2	108	Central wall of the vessel
K46	0.5	11	22	2	133	Central wall of the vessel
K47	0.5	11	22	4	59	Central wall of the vessel
K48	0.5	11	22	6	59	Central wall of the vessel
K49	0.5	11	22	8	59	Central wall of the vessel
K50	2	22	11	2	59	Central wall of the vessel
K51	2.5	55	22	2	59	Central wall of the vessel
K52	5.1	113	22	2	59	Central wall of the vessel
K53	10.3	113	11	2	20	Central wall of the vessel
K54	10.3	113	11	2	59	Central wall of the vessel
K55	10.3	113	11	2	108	Central wall of the vessel
K56	10.3	113	11	2	133	Central wall of the vessel
K57	10.3	113	11	4	59	Central wall of the vessel
K58	10.3	113	11	6	59	Central wall of the vessel
K59	10.3	113	11	8	59	Central wall of the vessel

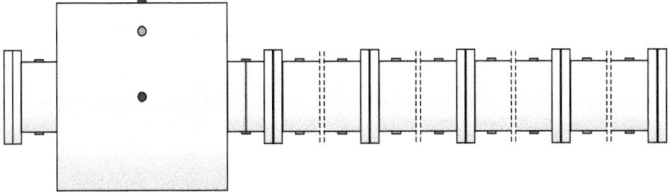

Fig. 4.3 Schematic of vessel–pipe structure

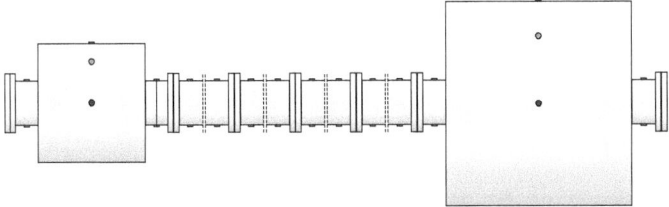

Fig. 4.4 Schematic of vessel–pipe–vessel structure

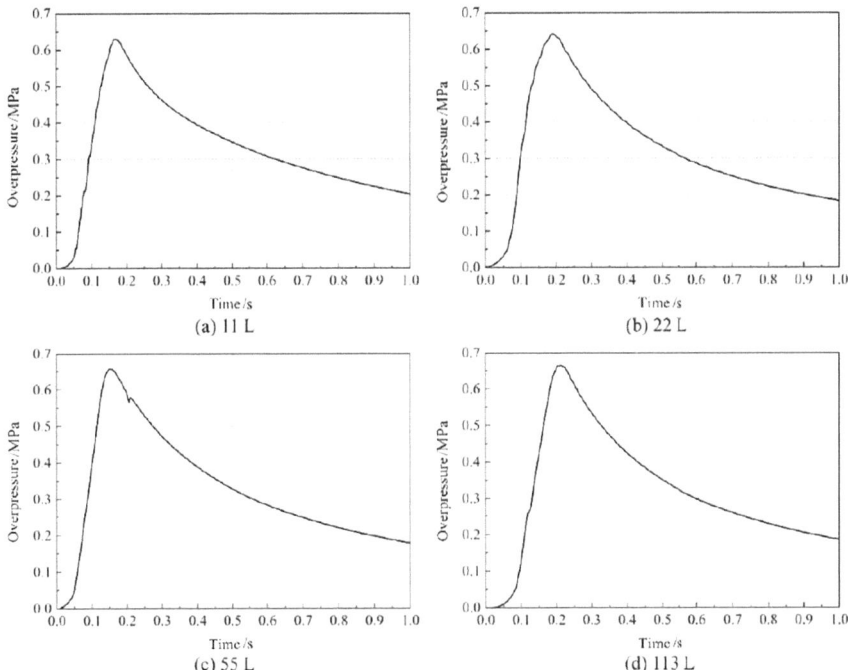

Fig. 4.5 Overpressure histories in the cylindrical vessel with different volumes

Table 4.4 Explosion
intensity (pressure) in a single
cylindrical vessel

Volume of vessel (L)	P_{max} (MPa)	$(dP/dt)_{max}$ (MPa s^{-1})
11	0.63	12.64
22	0.64	10.41
55	0.66	7.58
113	0.67	5.96

4.3 Scale Effect of Pipe–Vessel Structure

4.3.1 Effect of Pipe Length

In order to study the effect of pipe length on the explosion intensity of premixed methane–air in a vessel–pipe structure, this section utilized four types of single cylindrical vessels (11, 22, 55, and 113 L) connected by pipes of four different lengths (2, 4, 6, and 8 m) with the same inner diameter (59 mm). The ignition point was located at the center of the cylindrical vessel, and pressure sensors were positioned on the vessel wall and at the end of the pipe to collect data.

4.3.1.1 Overpressure in the Cylindrical Vessel

Figure 4.6 shows the variation of overpressure over time for premixed methane–air in 11 L (a), 22 L (b), 55 L (c), and 113 L (d) cylindrical vessels connected to pipes of different lengths. The maximum pressure (P_{max}) and maximum rate of pressure rise ($(dP/dt)_{max}$) were shown in Table 4.5. From Fig. 4.6 and Table 4.5, it was evident that the pipe length has a significant impact on the explosion characteristics within the cylindrical vessels. As the pipe length increased, P_{max} in all four vessels gradually decreased, while $((dP/dt)_{max})$ showed a trend of gradual increase.

In vessel–pipe structure, the change in pipe length made the scale effect of the pipe more pronounced. The increase in pipe length resulted in a larger contact area between the explosion flame and the pipe wall, leading to increased explosion energy loss and decreased P_{max} within the vessel [1–3]. Additionally, when the pipe length was increased, a greatest reduction in P_{max} occurred in the 11 L cylindrical vessel, while the reduction in the 113 L cylindrical vessel was relatively slower. This phenomenon can be attributed to the larger volume of the 113 L cylindrical vessel, which leads to a diminished pressure relief effect of the pipe on the P_{max} within the vessel [4–6].

According to the experimental results, in industrial production, connecting pipes of a certain length to pressure vessels can exert a pressure relief effect, which can effectively reduce the P_{max} in the pressure vessel following an explosion, thereby protecting the vessel.

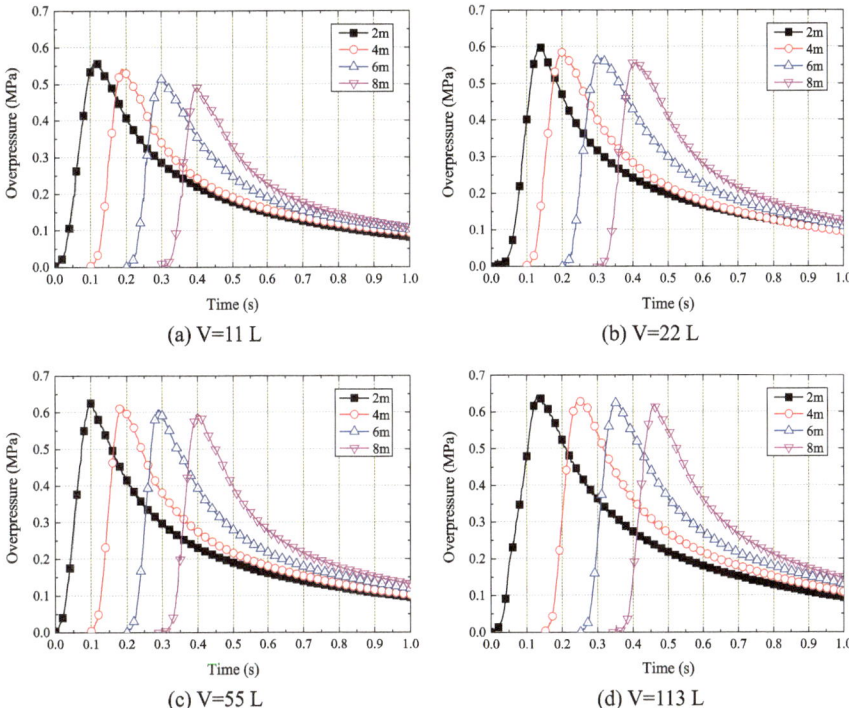

Fig. 4.6 Overpressure in cylindrical vessels with varying pipe lengths for different volumes

Table 4.5 Overpressure in cylindrical vessels connected to pipes of different pipe lengths

Volume of vessel (L)	Pipe length (m)	2 m	4 m	6 m	8 m
11	P_{max} (MPa)	0.56	0.54	0.51	0.48
	$(dP/dt)_{max}$ (MPa s^{-1})	16.25	18.28	20.86	23.89
22	P_{max} (MPa)	0.59	0.58	0.56	0.55
	$(dP/dt)_{max}$ (MPa s^{-1})	14.16	16.03	19.10	21.74
55	P_{max} (MPa)	0.62	0.61	0.60	0.59
	$(dP/dt)_{max}$ (MPa s^{-1})	12.12	14.06	16.42	19.47
113	P_{max} (MPa)	0.64	0.62	0.62	0.61
	$(dP/dt)_{max}$ (MPa s^{-1})	10.44	12.48	14.85	17.36

4.3.1.2   Overpressure at Pipe End

Figure 4.7 shows the overpressure histories at the end of the pipe 11, 22, 55, and 113 L cylindrical vessels connected to pipes of different lengths (2, 4, 6, and 8 m). P_{max} and $(dP/dt)_{max}$ were shown in Table 4.6. From Fig. 4.7 and Table 4.6, it can be seen that when a single cylindrical vessel was connected to a pipe with the same

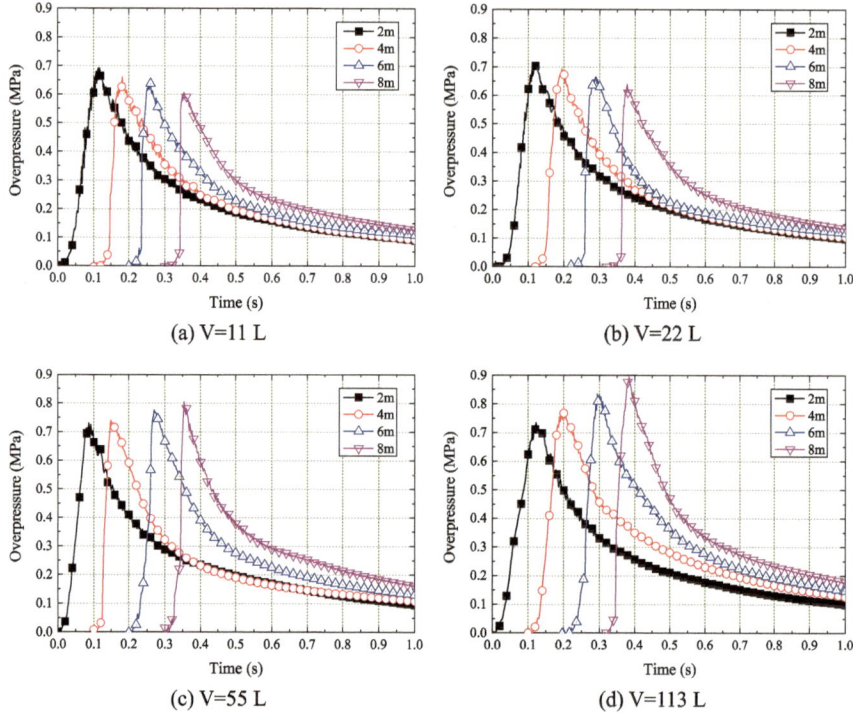

Fig. 4.7 Overpressure histories at the end of pipes with varying pipe lengths for different volumes

inner diameter but different lengths, the change in pipe length has a significant scale effect on the gas explosion at the end of the pipe. When the volume of cylindrical vessels were 11 and 22 L, the P_{max} at the end of the conduit gradually decreases with the increase of pipe length, while the $(dP/dt)_{max}$ gradually increases. However, for cylindrical vessels with larger volumes of 55 and 113 L, P_{max} and $(dP/dt)_{max}$ at the pipe end increase with the length of pipes. This was because the content of premixed methane–air in vessels with larger volumes also increases, resulting in a larger P_{max} generated after an explosion. The longer the pipe was, the more obvious the acceleration effect of the explosion flame, so the P_{max} also increases [7].

4.3.2 Effect of Pipe Inner Diameter

In order to analyze the effect of the inner diameter of the pipe on the explosion characteristics inside the cylindrical vessel and at the pipe end in a cylindrical vessel–pipe connection structure, four individual cylindrical vessels with different volumes (11, 22, 55, and 113 L) and four pipe connections with different inner diameters (20,

Table 4.6 Overpressure at the end of pipes with varying pipe lengths for different volumes

Volume of vessel V (L)	Pipe lengths L (m)	2 m	4 m	6 m	8 m
11	P_{max} (MPa)	0.69	0.65	0.63	0.60
	$(dP/dt)_{max}$ (MPa s^{-1})	27.42	47.57	89.95	127.41
22	P_{max} (MPa)	0.71	0.69	0.66	0.64
	$(dP/dt)_{max}$ (MPa s^{-1})	19.29	40.53	78.07	119.50
55	P_{max} (MPa)	0.72	0.73	0.77	0.80
	$(dP/dt)_{max}$ (MPa s^{-1})	16.15	32.59	66.00	99.62
113	P_{max} (MPa)	0.73	0.77	0.83	0.89
	$(dP/dt)_{max}$ (MPa s^{-1})	12.50	25.50	43.49	85.81

59, 108, and 133 mm) and the same length (2 m) were selected in this section, and the ignition position was located at the vessel.

4.3.2.1   Overpressure in the Cylindrical Vessel

Figure 4.8 shows the curves of overpressures of premixed methane–air in the cylindrical vessel as a function of time for 11, 22, 55 and 113 L cylindrical vessels connected to pipes of the same length (2 m) with different inner diameters (20, 59, 108, and 133 mm). The P_{max} and $(dP/dt)_{max}$ were shown in Table 4.7. It can be seen from Fig. 4.8 and Table 4.7 that the inner diameter of the pipe had a certain impact on gas explosion inside cylindrical vessel, and with the increase in the inner diameter of the pipe, the P_{max} and $(dP/dt)_{max}$ in the cylindrical vessel were increased.

4.3.2.2   Overpressure at Pipe End

Figure 4.9 shows the gas overpressure curves at the end of the pipe for different volumes of 11, 22, 55, and 113 L cylindrical vessels connected to pipes of different inner diameters. The maximum overpressure (P_{max}) and the maximum overpressure rise rate ($(dP/dt)_{max}$) was shown in Table 4.8. As can be seen from Fig. 4.9 and Table 4.8, the scale effect of gas explosion at the end of the pipe was observed for a single cylindrical vessel linked to a 2 m pipe with various inner diameters (20, 59, 108, and 133 mm). The maximum overpressure (P_{max}) and the maximum overpressure rise rate ($(dP/dt)_{max}$) at the end of the pipe exhibited an increase with the inner diameter of pipes for cylindrical vessels of 11 and 22 L. However, it was observed that the maximum overpressure (P_{max}) decreased with the inner diameter of the pipes, while the maximum overpressure rise rate ($(dP/dt)_{max}$) increased with the inner diameter of the pipes for cylindrical vessels of 55 and 113 L. This outcome mirrored the overpressure variations observed at the end of the pipe in the preceding section, indicating a correlation with the length-to-diameter (L/D) ratio of the pipe.

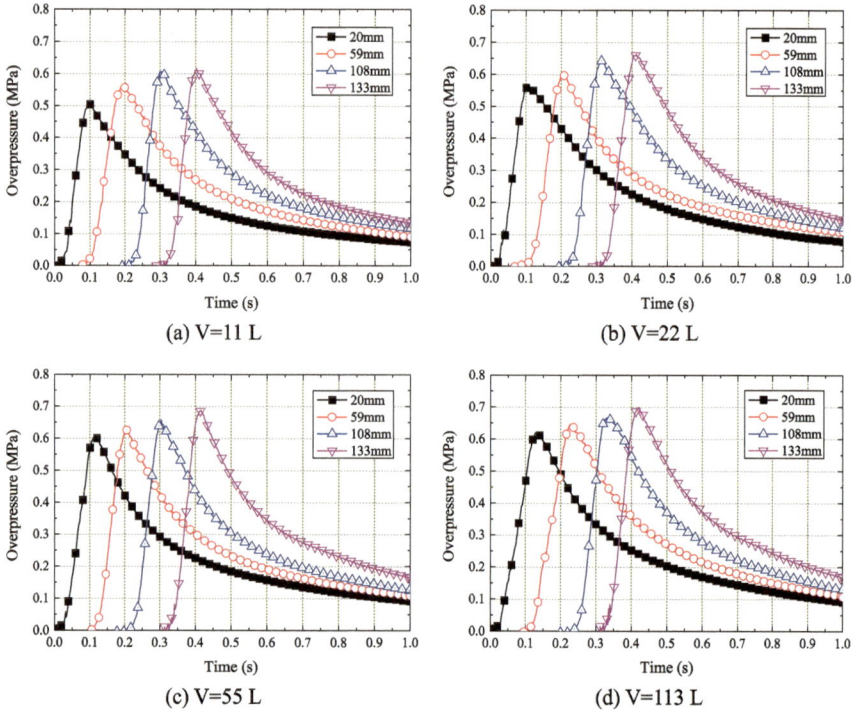

Fig. 4.8 Overpressure histories with pipes of different inner diameters for different vessels volumes

Table 4.7 Overpressure in cylindrical vessel after connecting different inner diameter pipes

Volume of vessel V (L)	Inner diameter of pipes D (mm)	20 mm	59 mm	108 mm	133 mm
11	P_{max} (MPa)	0.50	0.56	0.60	0.61
	$(dP/dt)_{max}$ (MPa s^{-1})	15.22	16.25	18.52	20.68
22	P_{max} (MPa)	0.56	0.59	0.64	0.66
	$(dP/dt)_{max}$ (MPa s^{-1})	13.01	14.16	16.97	18.55
55	P_{max} (MPa)	0.60	0.62	0.65	0.68
	$(dP/dt)_{max}$ (MPa s^{-1})	11.24	12.12	14.24	15.73
113	P_{max} (MPa)	0.61	0.64	0.66	0.70
	$(dP/dt)_{max}$ (MPa s^{-1})	9.27	10.44	12.65	13.28

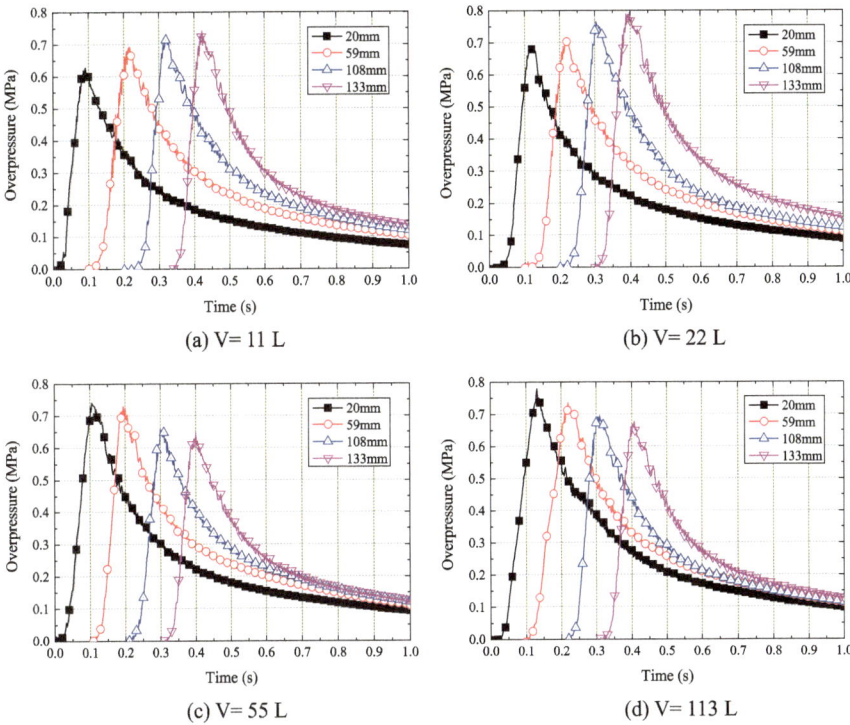

Fig. 4.9 Overpressure histories with different inner diameters at the pipe end

Table 4.8 Overpressure at the end of a vessel connected to pipes of different inner diameters

Volume of vessel V (L)	Inner diameter of pipes D (mm)	20 mm	59 mm	108 mm	133 mm
11	P_{max} (MPa)	0.62	0.69	0.72	0.73
	$(dP/dt)_{max}$ (MPa s^{-1})	25.86	27.42	29.76	32.14
22	P_{max} (MPa)	0.68	0.71	0.76	0.79
	$(dP/dt)_{max}$ (MPa s^{-1})	18.18	19.29	21.79	24.06
55	P_{max} (MPa)	0.74	0.72	0.65	0.63
	$(dP/dt)_{max}$ (MPa s^{-1})	14.49	16.15	17.09	21.36
113	P_{max} (MPa)	0.77	0.73	0.69	0.67
	$(dP/dt)_{max}$ (MPa s^{-1})	11.50	12.50	16.73	20.56

4.4 Scale Effect of Vessel–Pipe–Vessel Connected Structure

4.4.1 Effect of Volume Ratio

To investigate the effect of the volume ratio on the explosion characteristics within the vessel–pipe–vessel connected structure. In this study, four cylindrical vessels of varying volumes (11, 22, 55, 113 L) were matched with a pipe having a diameter of 59 mm and a length of 2 m. According to the experimental conditions, eight volume ratios were selected for analysis: 0.1, 0.2, 0.4, 0.5, 2.0, 2.5, 5.0, and 10.3. The overpressure histories under different volume ratios were shown in Figs. 4.10 and 4.11. The maximum overpressure (P_{max}) and the maximum overpressure rise rate (($dP/dt)_{max}$) was shown in Table 4.9.

The scale effect of volume ratios on gas explosions within the linked structural device was clearly observable. The maximum overpressure (P_{max}) and the maximum

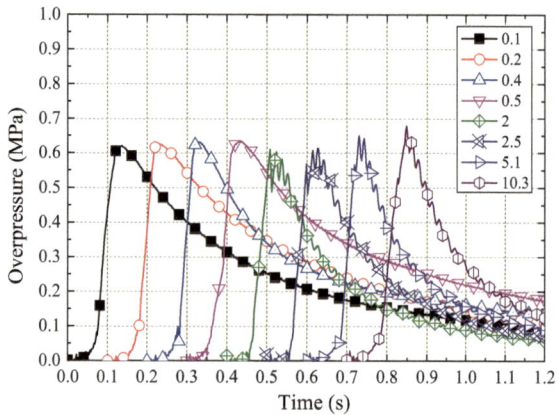

Fig. 4.10 Overpressure histories under different volume ratios at primary vessel

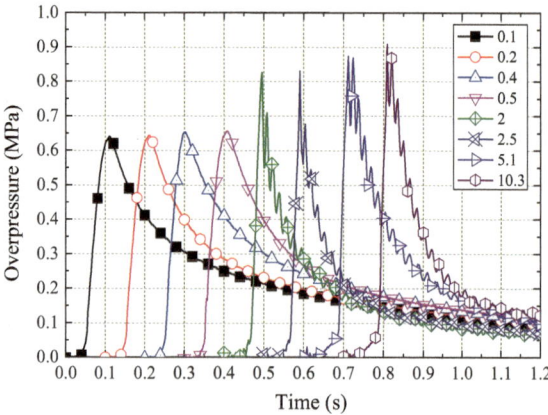

Fig. 4.11 Overpressure histories under different volume ratios at secondary vessel

Table 4.9 The explosion intensities of methane–air premixed gases in different volume ratios

	Volume ratio	0.1	0.2	0.4	0.5	2.0	2.5	5.0	10.3
Primary vessel	P_{max} (MPa)	0.61	0.62	0.63	0.63	0.60	0.61	0.64	0.67
	$(dP/dt)_{max}$ (MPa s^{-1})	17.16	20.36	22.70	24.72	22.30	20.24	17.06	15.09
Secondary vessel	P_{max} (MPa)	0.63	0.64	0.65	0.65	0.82	0.83	0.87	0.90
	$(dP/dt)_{max}$ (MPa s^{-1})	20.58	22.72	26.45	28.40	44.98	47.36	53.89	60.03

overpressure rise rate ($(dP/dt)_{max}$) increased with volume ratios in both the primary and secondary vessels when the volume ratios were less than 1. When the volume ratios exceed 1, the maximum overpressure (P_{max}) increased with volume ratios in both the primary and secondary vessels. Nevertheless, the maximum overpressure rise rate ($(dP/dt)_{max}$) increased with volume ratios in the primary vessel but decreases with volume ratios in the secondary vessel.

Meanwhile, there is no significant oscillation of the overpressure curves in both the primary and secondary vessels when the volume ratio was less than 1. Nevertheless, when the volume ratio was greater than 1, the overpressure curve exhibited pronounced oscillations in both the primary and secondary vessels, particularly after reaching P_{max}. Subsequently, the oscillations gradually subsided as the explosion advances.

Based on the experimental results, the following quantitative prediction model was obtained.

The prediction model of explosion intensity in the primary vessel when the volume ratio was less than 1 for methane–air premixed gases:

$$P_{max} = 0.61262 + 0.0706\lambda - 0.055\lambda^2, R^2 = 0.9775 \tag{4.3}$$

$$(dP/dt)_{max} = 14.72378 + 29.2519\lambda - 19.65\lambda^2, R^2 = 0.92022 \tag{4.4}$$

The prediction model of explosion intensity in secondary vessel when the volume ratio was less than 1 for methane–air premixed gases:

$$P_{max} = 0.63409 + 0.0541\lambda - 0.01667\lambda^2, R^2 = 0.97906 \tag{4.5}$$

$$(dP/dt)_{max} = 18.53324 + 21.2027\lambda - 3.045\lambda^2, R^2 = 0.99901 \tag{4.6}$$

The prediction model of explosion intensity in the primary vessel when the volume ratio was larger than 1 for methane–air premixed gases:

$$P_{max} = 0.57122 + 0.02013\lambda - 9.53681 \times 10^{-4}\lambda^2, R^2 = 0.98917 \tag{4.7}$$

$$(dP/dt)_{max} = 26.51988 - 2.64137\lambda + 0.14897\lambda^2, R^2 = 0.9411 \qquad (4.8)$$

The prediction model of explosion intensity in secondary vessel when the volume ratio was larger than 1 for methane–air premixed gases:

$$P_{max} = 0.78 + 0.02392\lambda - 0.00112\lambda^2, R^2 = 0.99404 \qquad (4.9)$$

$$(dP/dt)_{max} = 37.6662 + 4.21402\lambda - 0.19842\lambda^2, R^2 = 0.99411 \qquad (4.10)$$

where λ represented the volume ratios between the primary and the secondary vessels. All the fitted models exhibited confidence coefficients exceeding 0.9, affirming the predictive reliability.

4.4.2   Effect of Pipe Length

To investigate the effect of the pipe length on the explosion characteristics of the primary and secondary vessels within the vessel–pipe–vessel connected structure, two cylindrical vessels with volume ratios of 0.5 and 10.3 were paired with a pipe of 59 mm diameter, featuring varying pipe lengths of 2, 4, 6, and 8 m.

The overpressure histories under different pipe lengths for volume ratios of 0.5 and 10.3 were shown in Figs. 4.12 and 4.13, respectively. The maximum overpressure (P_{max}) and the maximum overpressure rising rate $((dP/dt)_{max})$ for volume ratios of 0.5 and 10.3 was shown in Tables 4.10 and 4.11, respectively.

It was evident from Figs. 4.12 and 4.13 that the scale impact of pipe lengths on gas explosions under volume ratios of 10.3 and 0.5 was clearly visible. The maximum overpressure (P_{max}) increased with pipe lengths in both the primary and

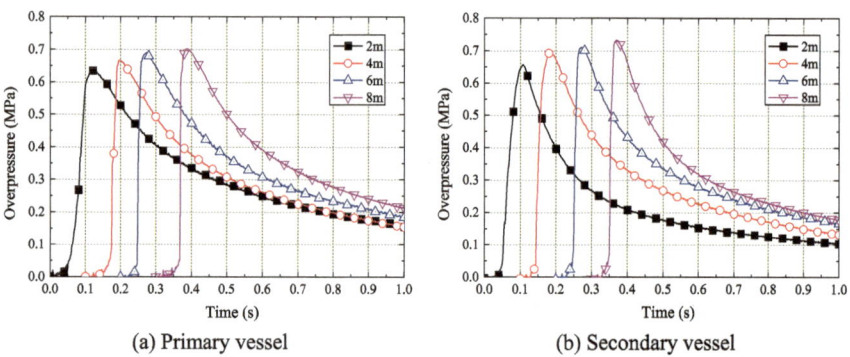

(a) Primary vessel   (b) Secondary vessel

Fig. 4.12 Overpressure histories under different pipe lengths for volume ratios 0.5

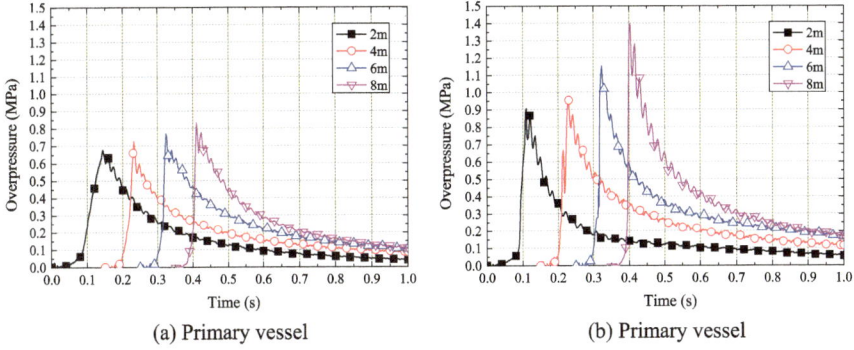

Fig. 4.13 Overpressure histories under different pipe lengths for volume ratios 10.3

Table 4.10 Overpressure in pipes of different lengths at the volume ratio of 0.5

Volume ratios 0.5	Pipe lengths L (m)	2 m	4 m	6 m	8 m
Primary vessel	P_{max} (MPa)	0.63	0.66	0.69	0.70
	$(dP/dt)_{max}$ (MPa s^{-1})	24.72	67.65	103.55	143.55
Secondary vessel	P_{max} (MPa)	0.65	0.69	0.71	0.73
	$(dP/dt)_{max}$ (MPa s^{-1})	28.40	56.44	81.40	104.13

Table 4.11 Overpressure pipes of different lengths at the volume ratio of 10.3

Volume ratios 10.3	Pipe lengths L (m)	2 m	4 m	6 m	8 m
Primary vessel	P_{max} (MPa)	0.67	0.72	0.77	0.83
	$(dP/dt)_{max}$ (MPa s^{-1})	15.09	42.30	76.22	102.31
Secondary vessel	P_{max} (MPa)	0.90	0.96	1.15	1.39
	$(dP/dt)_{max}$ (MPa s^{-1})	60.03	90.93	161.46	236.69

secondary vessels for both volume ratios. In contrast, the oscillation of the overpressure curve was minimal when the volume ratio was 0.5, whereas it becomes more pronounced with a volume ratio of 10.3. This overpressure oscillation phenomenon became increasingly noticeable with longer pipe lengths.

Table 4.10 shows that as the pipe length increases for 0.5 volume ratios, the maximum overpressure (P_{max}) increased from 0.63 to 0.70 MPa and the maximum overpressure rise rate ($(dP/dt)_{max}$) increased from 24.73 to 143.55 MPa s^{-1} at the primary vessel. The maximum overpressure (P_{max}) increased from 0.65 to 0.73 MPa and the maximum overpressure rise rate ($(dP/dt)_{max}$) increased from 28.40 to 104.13 MPa s^{-1} at the secondary vessel. The maximum increase of (P_{max}) and ($(dP/dt)_{max}$) was 0.03 MPa and 39.41 MPa s^{-1} for primary and secondary vessel, respectively.

It can be found from Table 4.10 that as the pipe length increases for 10.3 volume ratios, the maximum overpressure (P_{max}) increases from 0.68 to 0.83 MPa and the maximum overpressure rise rate (($dP/dt)_{max}$) increases from 15.09 to 102.31 MPa·s^{-1} at the primary vessel. The maximum overpressure (P_{max}) increased from 0.90 to 1.39 MPa and the maximum overpressure rise rate (($dP/dt)_{max}$) increased from 60.03 to 236.69 MPa s^{-1} at the secondary vessel. The maximum increase of (P_{max}) and (($dP/dt)_{max}$) was 0.56 MPa and 134.37 MPa s^{-1} for primary and secondary vessel, respectively.

As discussed above, there was a more obvious increase in overpressure for a larger volume, which was more dangerous.

Based on the experimental results, the prediction models of P_{max}, ($dP/dt)_{max}$ were obtained.

The prediction model of explosion intensity in the primary vessel when the volume ratio was 0.5 for methane–air premixed gases:

$$P_{max} = 0.59178 + 0.2377 \times l - 0.00123 \times l^2, R^2 = 0.99971 \tag{4.11}$$

$$(dP/dt)_{max} = -16.87828 + 21.44731 \times l - 0.18287 \times l^2, R^2 = 0.99758 \tag{4.12}$$

The prediction model of explosion intensity in the secondary vessel when the volume ratio was 0.5 for methane–air premixed gases:

$$P_{max} = 0.61557 + 0.02368 \times l - 0.00114 \times l^2, R^2 = 0.94056 \tag{4.13}$$

$$(dP/dt)_{max} = -2.0686 + 15.92222 \times l - 0.33145 \times l^2, R^2 = 0.99997 \tag{4.14}$$

The prediction model of explosion intensity in the primary vessel when the volume ratio was 10.3 for methane–air premixed gases:

$$P_{max} = 0.63978 + 0.01774 \times l + 7.5625 \times 10^{-4} \times l^2, R^2 = 0.99547 \tag{4.15}$$

$$(dP/dt)_{max} = -16.29452 + 15.47227 \times l - 0.06933 \times l^2, R^2 = 0.99276 \tag{4.16}$$

The prediction model of explosion intensity in the secondary vessel when the volume ratio was 10.3 for methane–air premixed gases:

$$P_{max} = 0.92595 - 0.03464 \times l + 0.01176 \times l^2, R^2 = 0.99324 \tag{4.17}$$

$$(dP/dt)_{max} = 42.56353 + 2.31987 \times l + 2.77059 \times l^2, R^2 = 0.99015 \tag{4.18}$$

where l (m) represents the pipe lengths, the confidence coefficients exceeding 0.9 affirm the predictive reliability.

It clear that the explosion intensity increased with pipe lengths in both the primary and secondary vessels. In particular, the overpressure occurred at the volume ratio of 10.3 for the secondary vessel. This was because the accelerating effect of the pipe on the explosion flame and pressure wave was more pronounced due to the increased pipe lengths, resulting in a higher reaction rate and greater turbulence intensity. Consequently, the explosion intensity of the methane–air premixed gas increased as the pipe lengths extends.

4.4.3   Effect of Pipe Inner Diameter

To investigate the effect of the inner diameter of pipe on the explosion characteristics of the primary and secondary vessels within the vessel–conduit–vessel connected structure, two cylindrical vessels with volume ratios of 0.5 and 10.3 were paired with a pipe length of 2 m, featuring varying inner diameter of pipe with 20, 59, 108 and 133 mm.

The overpressure histories for different inner diameters of the pipe at volume ratios of 0.5 and 10.3 were depicted in Figs. 4.14 and 4.15, respectively. The maximum pressure (P_{max}) and the maximum pressure rise rate (($dP/dt)_{max}$) for volume ratios of 0.5 and 10.3 under various inner diameters of the pipe were presented in Tables 4.12 and 4.13, respectively.

It was evident from Figs. 4.14 and 4.15 that the scale impact of inner diameters of the pipe on gas explosions under volume ratios of 10.3 and 0.5 was clearly visible. The maximum overpressure (P_{max}) decreased with the pipe inner diameters in both the primary and secondary vessels for both volume ratios of 10.3 and 0.5. Moreover, the pressure oscillation phenomenon was more pronounced for the volume ratio of 10.3 in comparison to the volume ratio of 0.5. The most pronounced oscillation occurred at the 20 mm inner pipe diameter, where P_{max} also increased from 0.70 to 0.94 MPa.

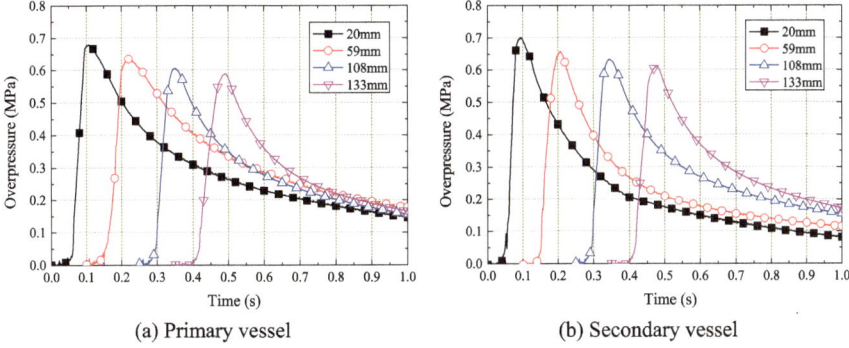

(a) Primary vessel   (b) Secondary vessel

Fig. 4.14 Overpressure histories with pipes of different internal diameters for volume ratios of 0.5

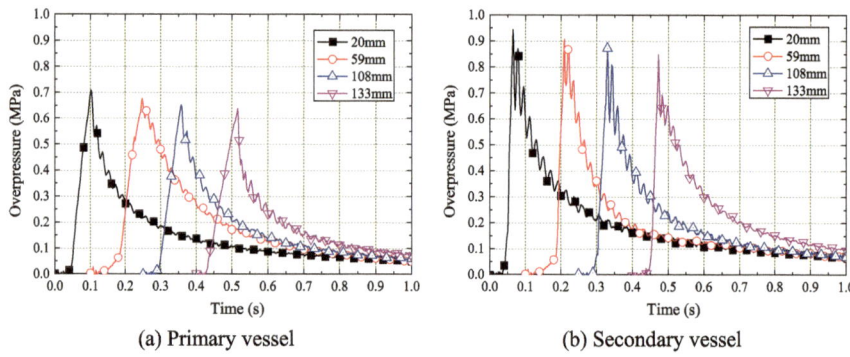

Fig. 4.15 Overpressure histories with pipes of different internal diameter for volume ratios of 10.3

Table 4.12 Overpressure with pipes of different internal diameter for volume ratios of 0.5

Volume ratios 0.5	Pipe lengths L (m)	20 mm	59 mm	108 mm	133 mm
Primary vessel	P_{max} (MPa)	0.67	0.63	0.60	0.58
	$(dP/dt)_{max}$ (MPa s^{-1})	28.85	24.72	19.89	16.92
Secondary vessel	P_{max} (MPa)	0.70	0.65	0.63	0.61
	$(dP/dt)_{max}$ (MPa s^{-1})	32.51	28.40	23.10	19.13

Table 4.13 Overpressure with pipes of different internal diameter for volume ratios of 10.3

Volume ratios 10.3	Pipe lengths L (m)	20 mm	59 mm	108 mm	133 mm
Primary vessel	P_{max} (MPa)	0.70	0.67	0.65	0.63
	$(dP/dt)_{max}$ (MPa s^{-1})	17.60	15.09	12.49	10.05
Secondary vessel	P_{max} (MPa)	0.94	0.90	0.87	0.84
	$(dP/dt)_{max}$ (MPa s^{-1})	74.11	60.03	52.87	46.76

From Tables 4.12 and 4.13, it was evident that the maximum overpressure rise rate $((dP/dt)_{max})$ decreases with the inner diameter of pipe in both the primary and secondary vessels for volume ratios of 10.3 and 0.5. This reduces the risk of explosion within the linked structure, and the minimized risk was achieved at 133 mm inner diameter. Based on the analysis above, it can be concluded that the scale effect of the inner diameter of the pipe was more significant at 10.3 volume ratios in comparison to 0.5 volume ratios. Therefore, the explosion hazard was less for larger pipe diameter in the linked structure.

Based on the experiment result, the prediction model of P_{max}, $(dP/dt)_{max}$ can be obtained.

The prediction model of explosion intensity in the primary vessel when the volume ratio was 0.5 for methane–air premixed gases:

$$P_{max} = 0.70361 - 0.00133 \times d + 3.57356 \times 10^{-6} \times d^2, R^2 = 0.98162 \quad (4.19)$$

$$(dP/dt)_{max} = 30.77257 - 0.09761 \times d - 4.35831 \times d^2, R^2 = 0.99785 \quad (4.20)$$

The prediction model of explosion intensity in secondary vessel when the volume ratio was 0.5 for methane–air premixed gases:

$$P_{max} = 0.72211 - 0.0012 \times d + 2.93706 \times 10^{-6} \times d^2, R^2 = 0.95178 \quad (4.21)$$

$$(dP/dt)_{max} = 34.04322 - 0.07578 \times d - 2.63747 \times 10^{-4} \times d^2, R^2 = 0.99422 \quad (4.22)$$

The prediction model of explosion intensity in the primary vessel when the volume ratio was 10.3 for methane–air premixed gases:

$$P_{max} = 0.72265 - 8.48115 \times 10^{-4} \times d + 1.68748 \times 10^{-6} \times d^2, R^2 = 0.99476 \quad (4.23)$$

$$(dP/dt)_{max} = 18.44002 - 0.04414 \times d - 1.31811 \times 10^{-4} \times d^2, R^2 = 0.97608 \quad (4.24)$$

The prediction model of explosion intensity in secondary vessel when the volume ratio was 10.3 for methane–air premixed gases:

$$P_{max} = 0.95792 - 8.20475 \times 10^{-4} \times d + 1.01892 \times 10^{-7} \times d^2, R^2 = 0.99147 \quad (4.25)$$

$$(dP/dt)_{max} = 81.13481 - 0.39916 \times d + 0.00111 \times d^2, R^2 = 0.95143 \quad (4.26)$$

where d (m) represented the inner diameter, the confidence coefficients exceeding 0.9 affirm the predictive reliability.

References

1. Zhao, H. (1996). *Principles of gas and dust explosions*. Beijing Institute of Technology Press.
2. Eisner, H. S. (1981). *Explosions: Course, prevention, protection* (W. Bartknecht, Trans.). Springer-Verlag.
3. Yin, Z., Wang, Z., Zhen, Y., et al. (2021). Propagation characteristics of gas explosion in linked vessels based on DDT criteria. *Journal of Loss Prevention in the Process Industries, 73*, 104598.
4. Xu, J., Zhou, X., & Wu, B. (2001). Study on the size effect of gas explosion propagation in mines. *Journal of Safety Science and Technology, 11*(6), 5.
5. Wei, C. (2010). *Research on the explosion process of methane-air premixed gas in horizontal pipelines* [Master's thesis]. North University of China.

6. Zhu, C. J., Lin, B. Q., Jiang, B. Y., et al. (2013). Numerical simulation of blast wave oscillation effects on a premixed methane/air explosion in closed-end ducts. *Journal of Loss Prevention in the Process Industries, 26*(4), 851–861.
7. Kindracki, J., Kobiera, A., Rarata, G., et al. (2007). Influence of ignition position and obstacles on explosion development in methane–air mixture in closed vessels. *Journal of Loss Prevention in the Process Industries, 20*(4–6), 551–561.

Chapter 5
DDT Propagation Characteristics of Gas Explosion in the Confined Spaces

5.1 DDT Propagation Characteristics Inside the Linked Vessel

5.1.1 Experimental Apparatus and Methods

The experimental system included vessel–pipe, gas distribution, data acquisition, and ignition systems. Cylindrical vessels with equal length–diameter ratio were used in the experiment and their volumes were 113 and 11 L. The length of the pipe was 2 m and the diameter of the pipe was 60 mm, with totaling four sections. All these components were connected by flanges. Holes inside the vessels and pipes were used to install pressure transmitters, flame sensors and ignition devices, which could be connected according to the experimental needs. As shown in Figs. 5.1, 5.2 and 5.3, pipes, a vessel connected with pipes, and two vessels connected with pipes were selected; P in the figure represents a pressure sensor, and a flame transducer. The distance between the measuring points was 1200 mm.

The ignition system was constructed using a high-energy electronic ignition device (XDH-6L). The range of ignition energy was 0–5 J. The ignition position was on the wall of 113 L cylindrical vessel or the left of the single pipe. A SY-9560 gas sample compounder was used to obtain the methane–air mixture. A high-frequency pressure transmitter (HM90H3-2) was used to measure the overpressure inside the vessels. The pressure transmitter range was 0–5 MPa. The frequency response was 200 kHz. A CKG-100 flame sensor was adopted to measure the duration of the flame. The pressure and temperature transducers were flush-mounted. A DEWE-16 multichannel data acquisition instrument was used to match with the pressure transmitters and flame sensors. The sampling rate of the data acquisition instrument was 200 kHz. The vessels and pipes were vacuumed to 0.01 MPa and filled with the stoichiometric concentration of methane–air mixture at 10% by the computer automatic gas distribution system.

© The Author(s) 2025
Z. Wang and X. Cao, *Gas Explosion and Its Protection Technology in Process Industries*, https://doi.org/10.1007/978-981-96-3121-6_5

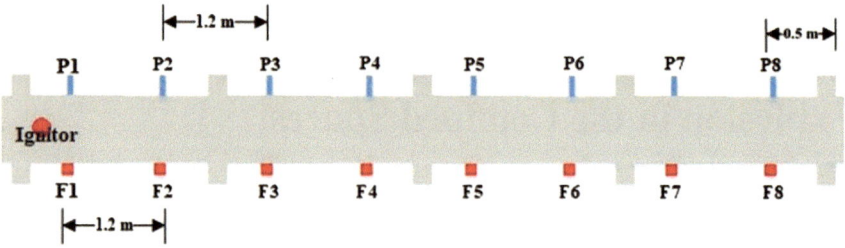

Fig. 5.1 Schematics of pipes

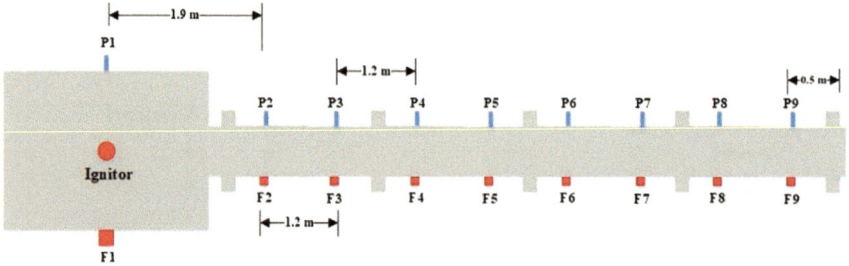

Fig. 5.2 Schematics of vessel connected with pipes

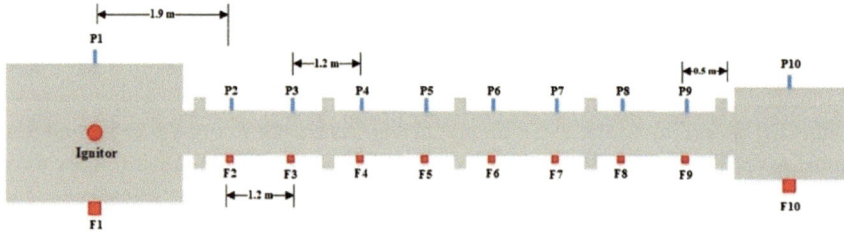

Fig. 5.3 Schematics of two vessels connected with pipes

In an experiment, the cylindrical vessels and pipes were connected by bolts and flanges. The vessels and pipe were vacuumed to 0.01 MPa and filled with 10% methane concentration by the computer automatic gas distribution system. Then, the ignition electrode mounted on the cylindrical vessel was discharged to generate electric spark and ignite the premixed gas. At the same time, the recorder was triggered to collect the signals of each pressure transmitter and flame sensor. After the explosion, the combustion products were purged with fresh air by a compressor to avoid the interference to next experiment.

5.1.2   Numerical Simulation Model

5.1.2.1   Establishment of Numerical Model

FLUENT simulation software was used in this study. Considering the time cost and axisymmetric structure of the pipes and linked vessels, a two-dimensional calculation was performed on the device (three-dimensional structure). A two-dimensional model for the pipes and linked vessels was selected. In the modeling, a fluid–thermal–solid interaction was set between the fluid and solid regions. The effects of heat transfer, thermal radiation, and gravity were considered.

5.1.2.2   Computational Domain and Grid Partition

Considering the calculation convergence characteristic of the numerical simulation and the accuracy of the simulation results, a tetrahedral mesh was adopted. The time step was set to 10^{-5} s considering the accuracy and calculation time. When deflagration was simulated in the linked vessels, the grid was set to 5 mm. Meanwhile, the grid was set to 3 mm considering the special deflagration-to-detonation state. The encryption of the grid enabled the DDT to be captured.

5.1.2.3   Numerical Simulation Method

In this study, the k–ε model was adopted to describe the turbulence development during the explosion of the methane–air mixture. The EDC model was adopted to simulate the chemical reaction mechanism of gas combustion and explosion with high accuracy. Combined with the two factors of calculation stability and accuracy of calculation results, the transient term was calculated by the first-order implicit method, and the SIMPLE algorithm was used in pressure–velocity coupling. The density, energy, variable in reaction process, continuous, and dynamic equations were solved through a second-order upwind. A pressure correction equation was solved using the standard discrete analysis method.

5.1.2.4   Initial and Ignition Conditions

A 10% methane–air mixture was selected as the premixed gas. The initial time was t_0, the initial temperature was set to 300 K, and the initial pressure was 0.08 MPa. A range of 1/10 in the center of the vessels was set as the ignition area with a temperature of 2000 K to initiate a gas explosion.

5.1.2.5 Validation of Effectiveness

To form linked vessels, 113 and 11 L cylindrical vessels were selected, and the two vessels were connected to an 8 m pipe. The vessels were vacuumed and filled with a 10% premixed methane–air mixture. The overpressure inside the vessels was monitored. The comparison between the experimental and simulated overpressure history curves are shown in Fig. 5.4.

The results in Fig. 5.4 show that the overpressure trends are the same. The deviation of peak pressure obtained experimentally and numerically was 8.3%. It is reported that the deviation between the experimental and numerical simulation results should not exceed 10% [1]. Therefore, the numerical model established in this study is effective.

5.1.3 Criteria for DDT in Linked Vessels

For methane–air mixtures to transition from deflagration to detonation, three main criteria must be satisfied. The first was the CJ pressure criterion. The overpressure was recorded to determine whether the overpressure has reached or exceeded the CJ detonation pressure. The second was the CJ speed criterion. The pressure propagation was recorded to decide whether the wave front propagation velocity has reached the sound velocity of the combustion product, the steady explosion velocity during the experiment, or the CJ detonation velocity. The third was the energy sudden criterion. The pressure wave curve and flame propagation velocity are recorded to determine whether the overpressure and flame propagation velocity change in magnitude. If the three conditions above were satisfied, then the DDT will occur. At ambient temperature, the CJ detonation pressure and CJ detonation velocity calculated theoretically at 9.5% concentration were 1.86 MPa and 1987.40 m/s.

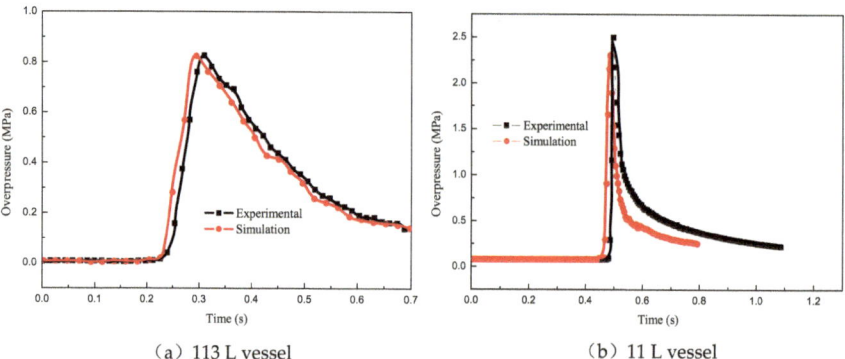

(a) 113 L vessel   (b) 11 L vessel

Fig. 5.4 Comparison between experimental and numerical simulation results

5.1.4 DDT Occurrence Condition in Different Structure

5.1.4.1 Pipe Structure

A 10% concentration of methane–air mixture was selected. For circular pipes of diameter 60 mm, 4, 6, and 8 m pipes were selected for the experiment and 9, 10, and 12 m for numerical simulation. The ignition position was on the wall of the left pipe. The pressure value and average flame propagation velocity were measured. Tables 5.1 and 5.2 show the maximum overpressure, distance from the ignition point, and maximum flame propagation velocity in the pipe.

As shown in Table 5.1, with the increase in pipe length, the peak pressure in a single pipe increased. As shown in Table 5.2, the flame propagation velocity was

Table 5.1 Maximum overpressure and its distribution in the single pipe

No.	Connection mode	Pipe diameter (mm)	Position of pressure peak	Pressure peak (MPa)	Maximum flame propagation velocity (m/s)	DDT occurrence
1	4 m pipe	60	P4	0.607	200	No
2	6 m pipe	–	P6	0.951	590	No
3	8 m pipe	–	P8	1.053	780	No
4	9 m pipe	–	P9	1.496	1105	No
5	10 m pipe	–	P10	1.671	1550	No
6	12 m pipe	–	P11	1.864	1906	Yes

Table 5.2 Maximum overpressure and its distribution inside the vessel–pipe structure

No.	Connection mode	Pipe diameter (mm)	Position of pressure peak	Pressure peak (MPa)	Maximum velocity (m/s)	DDT occurred
1	113 L vessel and 4 m pipe	60	P7	0.707	422	No
2	113 L vessel and 6 m pipe	–	P9	1.153	1065	No
3	113 L vessel and 8 m pipe	–	P11	2.106	1950	Yes
4	113 L vessel and 10 m pipe	–	P13	2.053	1980	Yes
5	113 L vessel and 12 m pipe	–	P15	2.174	2000	Yes
6	113 L vessel and 12 m pipe	30	P14	2.108	1988	Yes
7	113 L vessel and 6 m pipe	90	P9	2.018	1944	Yes

low. When the pipe length is less than 8 m, the maximum overpressure increases less with the increase in pipe length. As the pipe length continues to increase, the peak pressure and flame propagation velocity increase with the coupling of the flame and the pressure wave. When the pipe length was 12 m, DDT occurred in the single pipe. Methane is a low-reactive substance, and the length–diameter ratio required for the DDT was larger than that of the active fuel (hydrogen, acetylene, etc.) because the required acceleration section (induced distance for DDT) of the low-reactive substance was long. However, at a certain pipe diameter, DDT occurred as the pipe length increased.

As shown in Table 5.2, for the circular pipe with a diameter of 60 mm, the pipe of the peak pressure in the linked vessels increased with the pipe length. When the vessel was connected to the 8 m pipe, the pressure peak could reach 2.106 MPa, implying that the DDT occurred. However, the peak pressure points appeared at the end of the pipe. Because the pressure wave compressed the front unburned gas during the propagation, when the flame reached the end of the pipe, the pressure in the zone was higher than the initial pressure. Therefore, the pressure at the end of the pipe was higher than that in other areas of the linked vessels. Moreover, when the pipe diameter was larger than a certain value, the pipe promoted an increase in overpressure. The pipe length required for the DDT inside the vessel–pipe structure was shorter than the pipe length in a single pipe. This was because inside the vessel–pipe structure, the methane–air mixture in the pipe was ignited by the flame from the explosion vessel at a higher temperature, pressure, and degree of turbulence. Compared with that in a single pipe, the required acceleration phase in linked vessels was shorter.

5.1.4.2   Vessel–Pipe Structure

The 113 L cylindrical vessel was selected as the explosion vessel. For circular pipes with a diameter of 60 mm, 4, 6, 8, 10, and 12 m pipes were selected to study the explosion process. To study the effect of boundary conditions on the DDT, numerical simulations were performed on 30 and 90 mm diameter pipes. Tables 5.3 and 5.4 show the maximum overpressure, distance from the ignition point, and maximum flame propagation velocity in the pipe.

The effect of pipe diameter on the DDT of the methane–air mixture in the linked vessels was significant. As shown in Table 5.3, with the increase in pipe diameter, the pipe length required to form the DDT was small. This was because, in terms of heat transfer, as the diameter of the pipe decreases, more heat of gas per unit volume was consumed on the wall of pipes. The pressure wave surface structure was affected by the wall resistance and heat conduction, and the heat and momentum are lost. The smaller the pipe diameter, the more significant was the wall loss effect. Therefore, a small-diameter pipe (30 mm diameter) requires a long acceleration section. In the design of chemical plants, the pipe length and diameter should be reduced.

For the circular pipe with a diameter of 60 mm, the propagation velocity of the explosion flame increased with the pipe length. When the pipe length is 4 m, the maximum flame propagation velocity was 420 m/s, and the form of explosion was

Table 5.3 Maximum overpressure and its distribution inside the vessel–pipe–vessel structure

No.	Connection mode	Pipe diameter (mm)	Position of pressure peak	Pressure peak (MPa)	Maximum velocity (m/s)	DDT occurred
1	113 L vessel–4 m pipe–11 L vessel	60	P8	1.025	562	No
2	113 L vessel–6 m pipe–11 L vessel	–	P10	1.883	1990	Yes
3	113 L vessel–6 m pipe–11 L vessel	–	P12	2.145	2005	Yes
4	113 L vessel–10 m pipe–11 L vessel	–	P14	2.213	2009	Yes
5	113 L vessel–12 m pipe–11 L vessel	–	P16	2.237	2010	Yes
6	113 L vessel–8 m pipe–11 L vessel	30	P12	1.844	1893	Yes
7	113 L vessel–4 m pipe–11 L vessel	90	P8	1.850	1890	Yes

Table 5.4 Maximum flame velocity inside the vessel–pipe structure

No.	Connection mode	Pipe diameter (mm)	Maximum velocity (m/s)	DDT occurred
1	113 L vessel and 4 m pipe	60	422	No
2	113 L vessel and 6 m pipe	1065	No	
3	113 L vessel and 8 m pipe	1950	Yes	
4	113 L vessel and 10 m pipe	1980	Yes	
5	113 L vessel and 12 m pipe	2000	Yes	
6	113 L vessel and 12 m pipe	30	1988	Yes
7	113 L vessel and 6 m pipe	90	1944	Yes

deflagration. For a pipe length range of 4–8 m, the effect of pipe length on the acceleration of flame and pressure wave was significant. However, when detonation is formed, the flame propagation velocity tends to be stable with the increase in pipe length. It can be concluded that the detonation process is more stable than the deflagration process. As shown in Table 5.4, when the pipe is 12 m with a diameter of 30 mm, the DDT phenomenon can be observed based on the flame propagation velocity. The DDT can occur when the pipe is 6 m long with 60 mm in diameter.

This shows that the pipe length required to form the DDT is small when the pipe diameter is large.

5.1.4.3 Vessel–Pipe–Vessel Structure

The 113 L cylindrical vessel was selected as the explosion vessel. Numerical simulations were performed on 30- and 90-mm-diameter pipes. Tables 5.5 and 5.6 show the maximum overpressure, distance from the ignition point, and maximum flame propagation velocity in the pipe.

As shown in Table 5.3, the maximum overpressure appears in the explosion vessel. When the pipe measures 6 m long and 60 mm in diameter, DDT occurs inside the vessel–pipe–vessel structure. The peak pressure in the explosion vessel was several times higher than that in the single vessel. With the acceleration of explosion flame by the pipe, a high overpressure will be generated when the unburned gas in the explosion vessel was ignited. Moreover, it was noteworthy that the pipe length required for the DDT was short inside the vessel–pipe–vessel structure. In industrial anti-explosion engineering, the pipe length should be shortened to the maximum to DDT during accidents.

In addition, compared with the vessel–pipe structure, the pipe length required to form the DDT was shorter inside the vessel–pipe–vessel structure. Because a pressure imbalance occurred between the two vessels after ignition, the turbulence intensity of the unburned gas increased. The flame burning area increased and the acceleration was evident. As shown in Table 5.5, with the increase in pipe diameter, the pipe length required to form the DDT was smaller. Under the same pipe diameter, the pipe length required to form the DDT was smaller than that of the vessel–pipe structure.

Table 5.5 Maximum overpressure and its distribution inside the vessel–pipe–vessel structure

No.	Connection mode	Pipe diameter (mm)	Position of pressure peak	Pressure peak (MPa)	DDT occurred
1	113 L vessel + 4 m pipe + 11 L vessel	60	P8	1.025	No
2	113 L vessel + 6 m pipe + 11 L vessel		P10	1.883	Yes
3	113 L vessel + 6 m pipe + 11 L vessel		P12	2.145	Yes
4	113 L vessel + 10 m pipe–11 L vessel		P14	2.213	Yes
5	113 L vessel + 12 m pipe + 11 L vessel		P16	2.237	Yes
6	113 L vessel + 8 m pipe + 11 L vessel	30	P12	1.844	Yes
7	113 L vessel + 4 m pipe + 11 L vessel	90	P8	1.850	Yes

Table 5.6 Maximum flame velocity inside the vessel–pipe–vessel structure

No.	Connection mode	Pipe diameter (mm)	Maximum flame propagation velocity (m/s)	DDT occurred
1	113 L vessel + 4 m pipe + 11 L vessel	60	562	No
2	113 L vessel + 6 m pipe + 11 L vessel		1990	Yes
3	113 L vessel + 6 m pipe + 11 L vessel		2005	Yes
4	113 L vessel + 10 m pipe–11 L vessel		2009	Yes
5	113 L vessel + 12 m pipe + 11 L vessel		2010	Yes
6	113 L vessel + 8 m pipe + 11 L vessel	30	1893	Yes
7	113 L vessel + 4 m pipe + 11 L vessel	90	1890	Yes

This further illustrates that the DDT is likely to form inside the vessel–pipe–vessel structure.

The DDT was formed on the pipe measuring 6 m long and 60 mm in diameter, and the flame acceleration effect is obvious before the DDT occurrence. With the increase of pipe length, the flame velocities were almost the same after DDT occurred in the linked vessels. It showed that the explosion process becomes stable. Therefore, with the increase in pipe length, the flame velocity tends to be stable. In addition, under the same pipe length, the maximum flame propagation velocity inside the vessel–pipe–vessel structure is higher than that inside the vessel–pipe structure. According to the CJ velocity criterion, the DDT can be formed on a pipe measuring 8 m long and 30 mm in diameter. Furthermore, the DDT can be formed on a pipe measuring 4 m long and 90 mm in diameter. For the vessel–pipe–vessel structure, the DDT can occur in a large-diameter pipe under the condition of a short pipe length. Therefore, this structure type should be avoided, or its safety protection must be enhanced.

5.1.5 DDT Propagation Characteristics in Linked Vessels

5.1.5.1 Propagation Characteristics in Vessel–Pipe Structure

The 113 L cylindrical vessel was selected as the explosion vessel to connect an 8 m circular pipe. Figure 5.5 shows the time history curves of overpressure for each pressure monitoring point. In Fig. 5.5, Position 1 (P1) is the monitoring point of overpressure in the cylindrical vessel. When the methane–air mixture is ignited, it

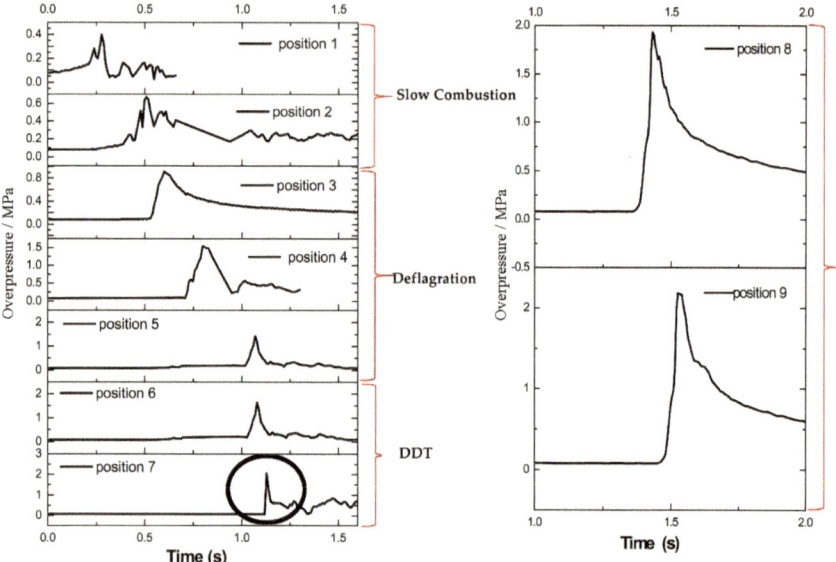

Fig. 5.5 Time history curves of overpressure for each pressure monitoring point under the structure of 113 L vessel connected to 8 m pipe

burns evenly from the ignition center to all sides. At this time, the overpressure increases slowly, and the pressure fluctuation is evident.

In Fig. 5.6, the flame propagation velocity shows a slow combustion. The unburned combustible mixture in the pipe affected by the explosion exhibits high turbulence, temperature, and pressure. When the flame is transmitted from the vessel to the pipe, the unburned combustible mixture in the pipe is ignited. Subsequently, the explosion enters the stage of rapid combustion. As shown in Fig. 5.5, the overpressure increases at P3 and P5. Figure 5.6 shows that the flame has entered the acceleration stage in the pipe, which shows that the pipe has a certain acceleration effect on the flame. The DDT was in the range of P6–P7. After reaching P7, the unburned gas explodes at the precursor shock wave. The form of explosion changes from deflagration to detonation. Subsequently, a detonation stage occurs at P8 and P9. At this stage, the pressure forms exhibit relatively stable characteristics. In addition, although the DDT appears in the later parts of the pipe, the overpressure of the cylindrical vessels was low because the pipes contribute to the pressure relief of the vessel. Additionally, vessels can be protected by connecting pressure relief ducts in industrial processes.

The flame velocity diagram is shown in Fig. 5.6. The explosion can be divided into four stages in this condition, i.e., slow combustion, deflagration, deflagration to detonation, and stable detonation. Figure 5.7 shows the development of the temperature field in the 8-m-vessel pipe structure. The flame acceleration was slow in the deflagration stage, and the flame propagation velocity increased rapidly when the DDT occurs.

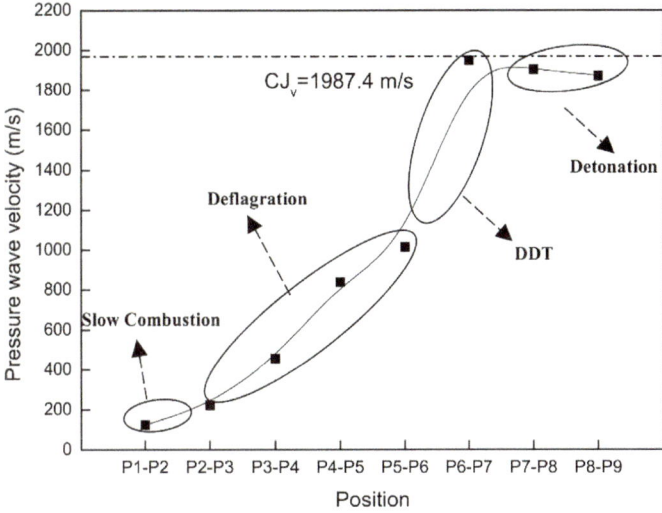

Fig. 5.6 Flame velocity diagram under the structure of 113 L vessel connected to 8 m pipe

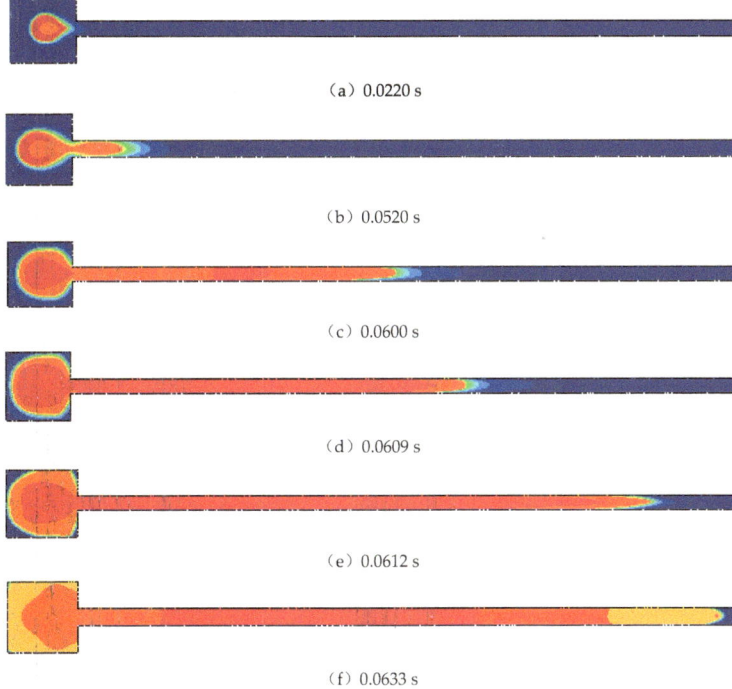

Fig. 5.7 Process map of temperature field development under the structure of 113 L vessel connected to 8 m pipe

5.1.5.2   Propagation Characteristics in Vessel–Pipe–Vessel Structure

The 113 L cylindrical vessel was selected as the explosion vessel, which was connected to the 11 L cylindrical vessel by an 8 m circular pipe. Figure 5.8 shows the time history curves of overpressure for each pressure monitoring point under the structure. As shown in Fig. 5.8, the pressure history curve at Position 7 (P7) increases significantly, which is consistent with the characteristic of the detonation pressure curve. The pressure peak at P7 is higher than those at P8 and P9. This indicates that local explosion at the precursor shock wave, i.e., the explosion process changes from deflagration to detonation and a strong detonation occurs. Because no obstacles exist in the circular pipe, a pressure decay occurs, and a stable detonation is formed after the strong detonation. The pressure peaks at P8 and P9 are near the theoretical CJ detonation pressure. As shown in Fig. 5.8, a strong detonation occurs at P10. Owing to the complexity of the vessel–pipe–vessel structure, an unburned gas at a high pressure is formed in the explosion vessel. Subsequently, a pre-compressed flammable gas is ignited by the jet flame at a high velocity, and a strong detonation occurs in the explosion vessel.

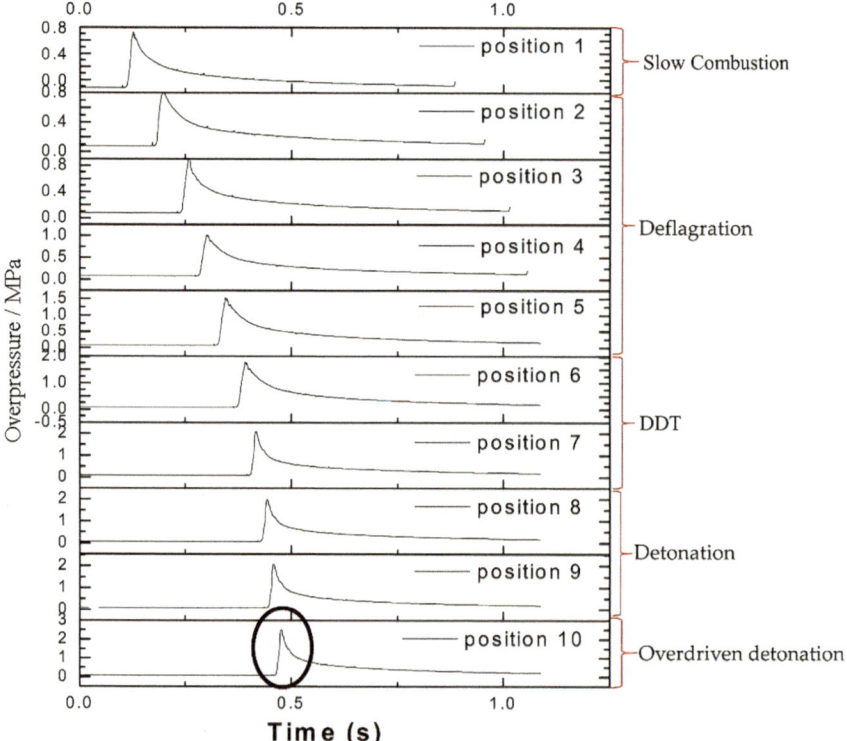

Fig. 5.8 Time history curves of overpressure for each pressure monitoring point under the structures of 113 and 11 L vessels connected to 8 m pipe

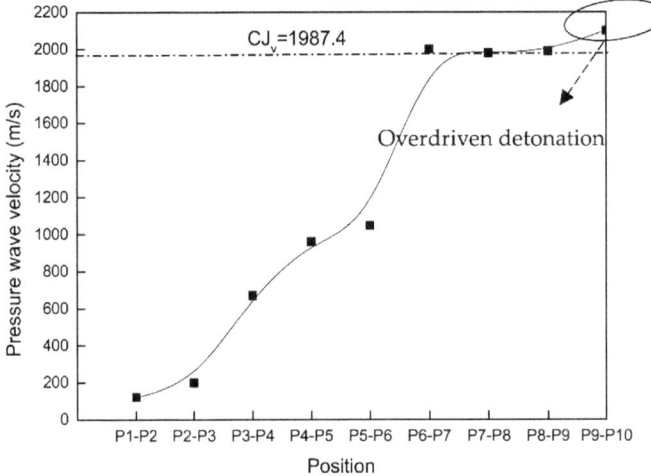

Fig. 5.9 Flame velocity under the structures of 113 and 11 L vessels connected to 8 m pipe

Figure 5.9 shows the flame velocity diagram of the linked vessels. The explosion wave propagation velocities at P6 and P7 are close to 2100 m/s, and the DDT occurs at this stage. Subsequently, the flame propagation velocity is relatively stable, and the flame speed finally reaches its maximum value inside the vessels (P9–P10). Before the detonation, the pressure peaks inside the vessel–pipe–vessel structure are higher than that inside the vessel–pipe vessel. The flame propagation process inside the vessel–pipe–vessel structure includes slow combustion, deflagration, deflagration to detonation, and detonation. However, compared with the situation inside the vessel–pipe structure, an overdriven detonation occurred inside the vessel–pipe–vessel structure. This phenomenon is related to the coupling among the precompression phenomenon inside the 11 L vessel, high turbulence, and jet flames from the pipe.

5.2 Effects of Obstacles on DDT in Linked Vessels

5.2.1 Experimental Apparatus and Methods

5.2.1.1 Experimental Apparatus

The experimental apparatus is shown in Fig. 5.10, including the vessel–pipe system, gas distribution system, data acquisition system and ignition system.

The experimental apparatus was made up of vessels and pipes. Cylindrical vessels with equal length–diameter ratio were selected and their volumes were 113 and 11 L. The length of the pipe was 2 m and the diameter of the pipe was 60 mm, totaling four sections. All these components were connected by flanges. Holes inside the

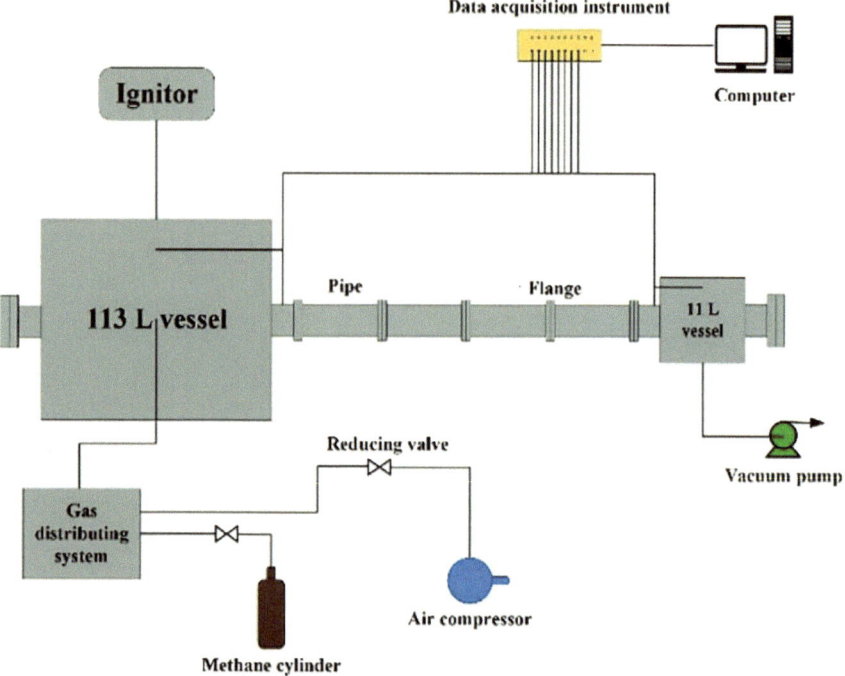

Fig. 5.10 Experimental system diagram of linked vessels

vessels and pipes were used to install pressure transmitters, flame sensors and ignition devices, which could be connected according to the experimental needs.

The ignition system was obtained by high energy electronic ignition device (XDH-6L). The ignition position was on the wall of 113 L cylindrical vessel. A SY-9560 gas sample compounder was used to obtain the mixture of methane and air of the desired concentration. The data acquisition system consisted of pressure transducer, flame sensor and data acquisition instrument. High-frequency pressure transmitter (HM90-H3-2) was used to measure overpressure inside the vessels. The pressure transducers are flush-mounted. A CKG-100 flame sensor was adopted to measure the duration of the flame. In order to prevent the high temperature in detonation and the impact of small particles, a layer of high vacuum silicone grease was coated on the surface of flame sensors. A DEWE-16 multi-channel data acquisition instrument was used to match with pressure transmitters and flame sensors. In the paper, the experiments were carried out at room temperature and the initial pressure of reactant was 0.08 MPa.

5.2.1.2 Experimental Methods

The distance between obstacles affects the formation of unburned gas vortexes. Obstacles with wide spacing will generate large vortices, which can increase the flame propagation velocity [2]. However, the increase of vortex kinetic energy and scale will reduce the transport velocity. An appropriate distance in between obstacles can produce the best scale vortices, which will make the flame acceleration effect most obvious. When the distance in between the two adjacent obstacles is twice longer than the diameter of pipe, it is conductive to the flame acceleration [3]. Therefore, the distance between the two obstacles is set to be 0.12 m. Obstacles with the same blockage ratio were fixed by struts and bolts, which could be combined to form different blockage ratios and number of obstacle groups. The left side of the obstacle group was a circular iron ring with an inner diameter of 60 mm and an outer diameter of 90 mm. The iron ring could be fixed between the flanges, and prevent the obstacle group from moving back and forth due to the shock wave in the explosion. In addition, it is convenient to remove the obstacle groups in the pipe after the experiment.

As is illustrated in Fig. 5.11, four obstacle groups were set up in the pipe. Take the vessel–pipe–vessel structure as an example. There were 10 pressure measuring points to record the change of pressure in the linked vessels. The first and tenth pressure measuring points were located on the wall of two cylindrical vessels. The pressure measuring point P2 and P3 were 1.9 m and 3.1 m away from the ignitor, respectively. The ninth pressure measuring point P9 was 0.5 m away from the bottom of the pipe, and the other pressure measuring points were evenly distributed on the horizontal pipe. Similarly, the same number of flame sensors was installed to record the pressure wave curve and flame propagation velocity during the explosion process.

The cylindrical vessels and pipes were connected by bolts and flanges, and the silicone gasket was clamped in the middle of the flanges. The vessels and pipe were vacuumed to 0.01 MPa and filled with the premixed gas of 10% methane concentration by the computer automatic gas distribution system. Through the ignition system, the ignition electrode mounted on the cylindrical vessel was discharged to generate electric spark and ignite the premixed gas. At the same time, the recorder was triggered to collect the signals of each pressure transmitter and flame sensor. In the experiment of different blockage ratio, the selected annular orifice plates were

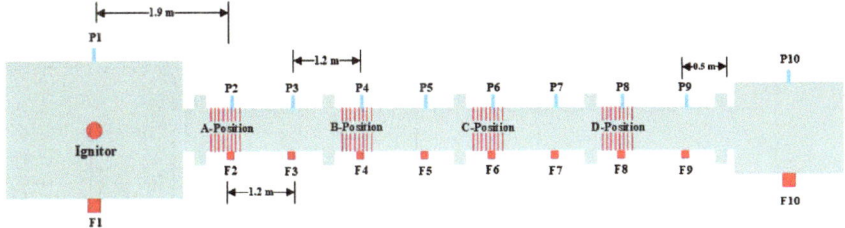

Fig. 5.11 The schematic diagram of linked vessel with obstacles

installed in the specified position and fixed. After the explosion, the combustion products were purged with fresh air by a compressor to avoid the interference to next experiment.

5.2.2 Typical DDT Process in Linked Vessels with Internal Obstacles

In this section, a 113 L cylindrical vessel was selected to connect an 8 m pipe, and 8 obstacles with 75% blockage ratio were installed at Position A. The ignition position is in the center of 113 L vessel. As shown in Fig. 5.12, we also calculated the pressure wave velocity when a 113 L cylindrical vessel was connected to an 8 m pipe, and 8 obstacles with 75% blockage ratio were installed at different positions. The maximum wave speeds in the pipe were 1923, 1814, 1909, and 1981 m/s when the obstacles were at Position A, B, C, and D, indicating that the detonation occurred in the pipes. It was also consistent with the method of pressure judgment, which proved that the pressure criterion was applicable.

Figure 5.13 shows the pressure histories in linked vessels for each pressure monitoring point. There are three main criteria for the DDT, including CJ pressure criterion, CJ speed criterion and energy sudden criterion [4–7]. The CJ pressure criterion is adopted in the paper. By theoretical calculation [8], the detonation pressure is 1.86 MPa under the experimental conditions. The detonation occurs when the overpressure reaches or exceeds the CJ pressure. The time shown in Fig. 5.13 is the relative time. There are obvious differences in trend between deflagration pressure history curves and detonation history curves. The deflagration pressure (P1, P2) rises slowly, and the peak pressure of deflagration is lower than that of detonation curve

Fig. 5.12 Pressure wave velocity distribution in linked vessels with obstacles

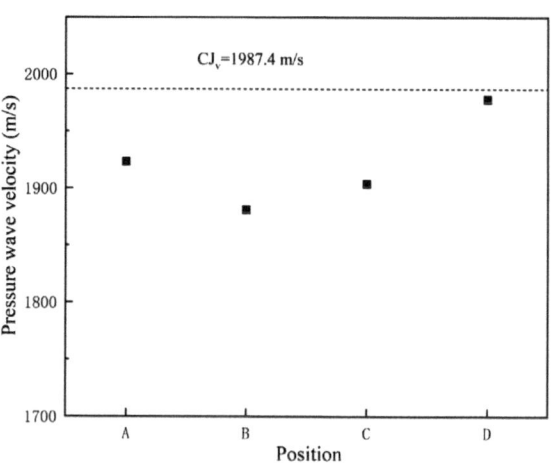

Fig. 5.13 Pressure histories
for each pressure monitoring
point

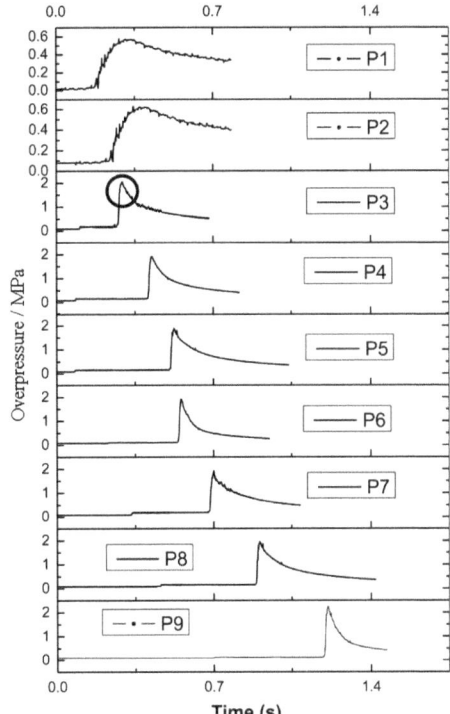

Fig. 5.13 Pressure histories for each pressure monitoring point

(P3–P9). The characteristics of detonation pressure were studied by taking P9 pressure history curve as an example. When the precursor shock wave passed through the measuring point P9, it caused a sudden jump in pressure, and then the pressure decreased owing to rarefaction wave [9].

When the vessel was connected to the pipe with obstacle group, the distance of DDT occurrence moved forward from P7 to P3. This indicates that the obstacle group has a high perturbation effect on the flame. Specifically, the flame propagation in the pipe with obstacles creates pockets of fresh fuel mixture between the obstacles. Gas expansion produces a strong jet flow in the unobstructed part of the pipe. The jet flow makes the flame tip to propagate much faster, which produces new pockets and leads to flame acceleration. DDT occurs at P3 under the coupling action of flame and pressure. As the combustion products expand, a compression wave was formed in front of the flame array. The faster the flame propagates, the sharper the area converges between the compression wave front and the flame surface. The overpressure increases accordingly. As can be seen from P4 and P5, after DDT occurred in the linked vessels, the peak pressures at the two measuring points were much the same. It showed that a steady detonation occurred in the pipe [10]. When the precursor shock wave propagated to the end of the pipe, the unburned gas mixture was compressed by the shock wave and affected by high temperature. The combustible gas exploded under high pressure and temperature, so the detonation pressure increased at P9.

To support the CJ pressure criterion, we also used the CJ speed criterion to evaluate the occurrence of detonation, taking P3 and the following measuring points as examples. The detonation reaction zone is accompanied by strong light, and the arrival time of the strong light can be detected by the flame sensor. Detonation propagation is the coupling propagation of detonation reaction zone and precursor shock wave. When the error is less than 10%, the coupled propagation of the precursor shock wave and the chemical reaction zone can be considered as detonation [11]. Figure 5.14 shows the relationship between detonation reaction zone arrival time and precursor shock time and the distance measured by flame sensor and pressure transducer. The time origin in Fig. 5.14 is the time when detonation occurs at P3 (T = 0 ms), and P3 is taken as the distance origin (L = 0 m). Figure 5.14 shows that the data error monitored by the flame sensor and the pressure transducer is less than 10%. By fitting the data obtained, the detonation velocity is 1923.0 m/s under operating conditions, which is 3% less compared with the theoretical value of 1987.4 m/s. Therefore, the process is taken as detonation propagation.

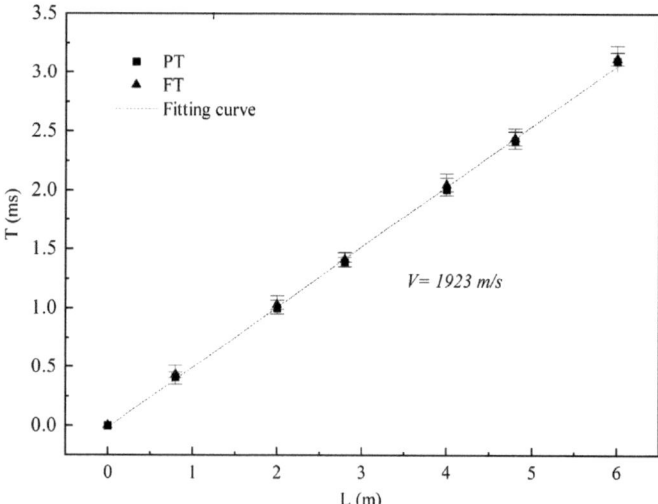

Fig. 5.14 The relationship between detonation reaction zone arrival time and precursor shock time and distance

5.2.3 Effects of Obstacles on DDT in Vessel–Pipe Structure

5.2.3.1 Obstacle Number

Obstacles with 75% blockage ratio were adopted and placed at Position A. The ignition position was on the wall of 113 L vessel. Pipes of 4, 6 and 8 m were selected to connect the vessel. Figure 5.15 shows the effect of obstacle number on DDT in vessel–pipe structure. The effect of obstacle number on the DDT is obvious. With the increase of pipe length, the overpressures decreased in the explosion vessel. The pipe has a pressure relief effect on the vessel.

Due to the change of the airflow propagation interface, the airflow is hindered, and the pressure at the entrance of the obstacles decreases significantly. While the flame overturns the obstacles, the increase of the flame surface accelerates the combustion. The deep and narrow space between the obstacles acts as a mini-channels, and the airflow will create turbulence in the pockets. Turbulence corrugates the flame front and increases the burning velocity. As the flame accelerates, compression waves and shock waves become stronger, until explosion starts and develops into detonation [12]. As a result, the turbulence of combustion is strengthened, and the peak pressure increases accordingly. Within the selected number of obstacles, the pressure peak increases with the increase in the number of obstacles. However, when DDT occurs in the linked vessels, the detonation pressure remains roughly unchanged. These further illustrate that a steady detonation occurred. Meanwhile, the mixed gas at the end of the pipe is affected by pre-compression and pre-radiation heat, so the pressure at the end of the pipe increases obviously. With the increase of the obstacle number, the distance of DDT occurrence becomes shorter.

Figure 5.15a shows that when the vessel is connected to a 4 m pipe, there is no DDT occurrence. The flame accelerates when passing through the obstacles, but meanwhile it is also hindered by the return gas from the end of the pipe. Thus, no steady detonation is formed when the pipe length is 4 m [13]. DDT can occur with the increase of obstacle number in this working condition. Figure 5.15b shows pressure peak distribution diagram of linked vessels when the vessel is connected to a 6 m pipe. When the number of obstacles is 6 or 8, there is DDT occurrence. DDT occurs at P6 with 6 obstacles, while DDT moves forward under eight obstacles and appears at P5. Figure 5.15c shows that DDT occurs at P3 under 8 obstacles when

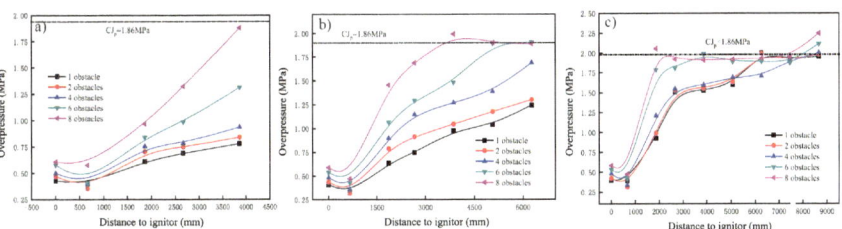

Fig. 5.15 Pressure histories of linked vessels with different number of obstacles

the vessel is connected to an 8 m pipe, and the detonation wave propagates in the pipe. This indicates that the detonation wave has the characteristic of self-sustaining propagation. Therefore, obstacles should be minimized to avoid the occurrence of detonation. For example, the corrosion of vessel wall should be reduced and the idle mechanical equipment in the roadway should be removed.

5.2.3.2   Obstacle Position

Eight obstacles with 75% blockage ratio were adopted. Figure 5.16 shows pressure history of the linked vessels with obstacles at different positions. The pressure behind the obstacle shows a higher growth ratio than that before the obstacle. When the flame passes through the obstacles group, the mixture of unburned combustible gas is disturbed to form turbulence, which will increase the flame area and thus accelerate the combustion ratio. This indicates that the position of the obstacle has a significant effect on DDT in linked vessels. For the 4 m pipe, the pressure distribution trends are the same at different obstacle positions, but the pressure peak at position B is lower than that at position A. When the obstacle is located at position B, the acceleration distance of high-speed flame is short after the flame passes through the obstacle. Therefore, the flame, passing through the obstacles, has spread to the end of the pipe before it accelerates. For the 6 m pipe, the pressure peaks at different locations are basically the same. As the obstacle approaches the ignition point, the peak pressure in the linked vessels increases. In addition, as the obstacle moves away from the ignition point, the distance of DDT occurrence becomes far away. Hence, anti-static and anti-spark measures should be taken near the obstacles to avoid the explosion accidents. For the 8 m pipe, DDT occurs first at position A. Due to the long pipe, the flame accelerates to a higher degree in the pipe, and the distance of DDT occurrence is relatively uniform when the obstacle is at positions B, C and D. Besides, after the detonation is formed, the obstacle has a further enhancement effect on the detonation wave. It can be seen from Fig. 5.16c that when the obstacle is at position D, the detonation pressure is larger at P7–P9 than that in other working conditions. In daily industrial production, when it is necessary to set obstacles such as instruments in the pipe, it is important to strengthen the safety technology and management measures near such obstacles.

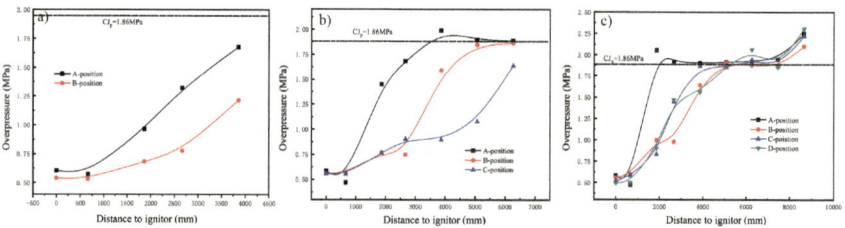

Fig. 5.16 Pressure histories of linked vessels with obstacles in different positions

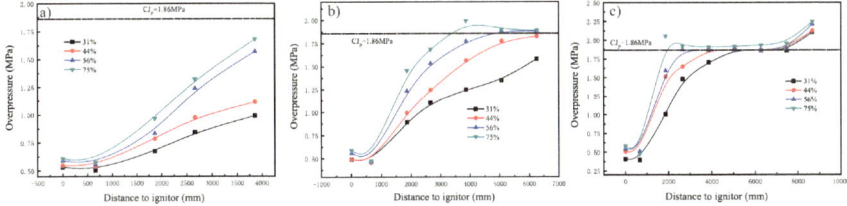

Fig. 5.17 Pressure histories of linked vessels with obstacles at different blockage ratios

5.2.3.3 Obstacle Blockage Ratio

Eight obstacles were adopted and the obstacles were placed at Position A. Figure 5.17 shows the pressure history of linked vessels with obstacles at different blockage ratios. From Fig. 5.17, it can be concluded that the blockage ratio has a significant effect on the explosion in linked vessels. The unburned combustible gas mixture in the obstacle group was disturbed and formed turbulence when the explosive flame passed through the obstacles. The turbulence then increased the flame area and accelerated the combustion rate, eventually leading to the increase of overpressure. In the present case, the blockage ratio plays an important role in the flame dynamics similar to the acceleration mechanism described by Bychkov [14]. The flame acceleration renders the compression wave stronger, until it develops into a shock of considerable amplitude. Compression of the fuel mixture by the shock is conventionally considered as one of the main elements of DDT in the pipe with obstacles. On the other hand, the obstacle blockage ratio is one of the decisive factors determining the degree of turbulence. The flame came into contact with the obstacle group when the blockage ratio increased to a certain value, and the heat loss dominated [15]. Thus, the turbulence effect of the flame was not enhanced but inhibited instead. For example, this mechanism was used for the inner core of the flame arrester to suppress the flame and even quench the effect.

5.2.4 Effects of Obstacles on DDT in Vessel–Pipe–Vessel Structure

5.2.4.1 Obstacle Number

Obstacles with 75% blockage ratio were used and the obstacle group was placed at Position A. The 113 L cylindrical vessel and the 11 L cylindrical vessel were connected by pipes. The ignition position was in the center of 113 L vessel. Pipes of 4, 6 and 8 m were selected to connect the vessels. Figure 5.18 shows the effect of obstacle number on DDT in vessel–pipe–vessel structure.

Figure 5.18 shows that as the number of obstacles increases, the pressure in the linked vessels all show an upward trend. With the same length of pipes, the pressure

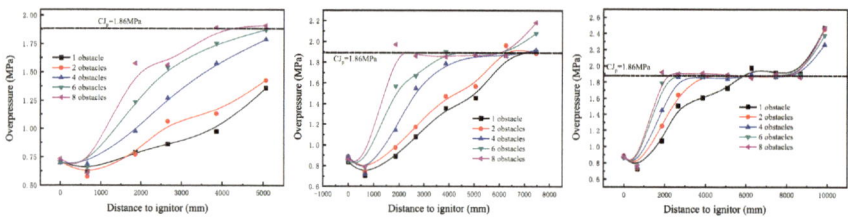

Fig. 5.18 Pressure histories of linked vessels with different number of obstacles

peak in the explosion vessel shows little change. This may be because the setting of obstacles increases the overpressure behind the obstacle. Moreover, the peak pressure at the entrance of the obstacles (P2) decreases because the airflow is blocked by the obstacles, which slows down the flame propagation velocity. The decrease of flame velocity reduces the energy of shock wave in the flame front, so the peak pressure decreases. When the number of obstacles increased to 6, DDT occurred in the linked vessels (4 m pipe). The addition of obstacle group in the linked vessels shortened the distance formed by deflagration to detonation transition. Figure 5.18b, c show the pressure history of linked vessels when 6 m pipe and 8 m pipe were linked to the vessels. Overdriven detonation occurred in 11 L vessel due to the coupling effects of high ignition pressure, high turbulence degree and jet flame.

5.2.4.2 Obstacle Position

The blockage ratio of obstacles was 75%, and the number of obstacles was 8. Pipes of 4, 6 and 8 m were selected to connect the vessels. Figure 5.19 shows the pressure history of each pressure sensor under different obstacle positions.

It can be seen from Fig. 5.19 that the position of the obstacle group affects the induction distance of DDT occurrence. Before the formation of detonation and at the same distance from the ignitor, the pressure behind the obstacle shows a higher growth ratio than that before the obstacle. At the entrance, the flow is hindered to some extent, so the pressure is reduced. When the flame passes through the obstacles, the turbulence of combustion is enhanced and results in the flame acceleration, so the pressure increases accordingly. For the two-vessel-connection structure, due to

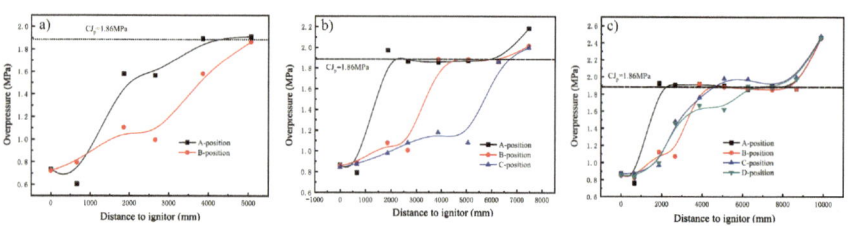

Fig. 5.19 Pressure histories of linked vessels with obstacles in different positions

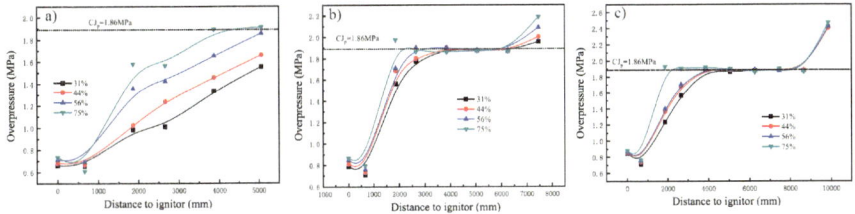

Fig. 5.20 Pressure histories of linked vessels with obstacles at different blockage ratio

its own pressure imbalance, the induction distance of DDT is shorter than that inside the vessel–pipe structure.

5.2.4.3   Obstacle Blockage Ratio

The number of obstacles was 8 and the obstacles were placed at Position A. Figure 5.20 shows the pressure history measured by each pressure sensor under different obstacle blockage ratios. In the range of selected blockage ratios, the overpressure rose as the blockage ratios of the obstacles increased.

Figure 5.20a shows that for a 4 m pipe, DDT occurs only when the obstacle blockage ratio is 75%. Lower obstacle blockage ratios have less effect on the degree of turbulence of the airflow. Hence, the flame cannot be accelerated to make DDT occur when the blockage ratio is low and the length of pipe is short. As can be seen from Fig. 5.20b, c, the DDT is more likely to occur with pipe length increased. Meanwhile, the blockage ratio of the obstacle has little effect on the detonation wave after the explosion develops into a steady detonation.

References

1. Feng, M. A. (2008). Simulation on pre-mixed methane-air explosion in the linked vessel with an inner obstacle. *Industrial Safety and Environmental Protection, 34*, 90–94.
2. Lu, J., Ning, J. G., Wang, C., et al. (2004). Study on flame propagation and acceleration mechanism of city coal gas. *Explosion and Shock Waves, 54*, 118–124.
3. Zipf, R. K., et al. (2014). Deflagration-to-detonation transition in natural gas–air mixtures. *Combustion and Flame, 161*(8), 2165–2176.
4. Aizawa, K., Yoshino, S., Mogi, T., et al. (2008). Study of detonation initiation in hydrogen/air flow. *Shock Waves, 18*(4), 299–305.
5. Duan, J. Y., Wang, J., & He, Z. (2010). Experimental study on the generating condition of over-detonation during the gaseous DDT. *Chinese Journal of High Pressure Physics, 24*(4), 305–310.
6. Lifshitz, A. (2001). *Handbook of shock waves.* Academic Press.
7. Rakotoarison, W., Pekalski, A., & Radulescu, M. I. (2020). Detonation transition criteria from the interaction of supersonic shock-flame complexes with different shaped obstacles. *Journal of Loss Prevention in the Process Industries, 64*, 103963.
8. Sun, J. S., & Zhu, J. S. (1995). *Theoretical explosion physics.* Ordnance Industry Press.

9. Yu, H. R., Chen, H., & Zhao, W. (2006). Advances in detonation driving techniques for a shock tube/tunnel. *Shock Waves, 15*(6), 399–405.
10. Huang, Z. P. (2006). *Explosion and shock measuring technique.* National Defense Industry Press.
11. Cox, G. (1995). *Combustion fundamentals of fire.* Academic Press.
12. Bychkov, V., Akkerman, V., Valiev, D., et al. (2010). Influence of gas compression on flame acceleration in channels with obstacles. *Combustion and Flame, 157*(10), 2008–2011.
13. Jin, T., Luo, K., Dai, Q., et al. (2016). Numerical study on three-dimensional CJ detonation waves interacting with isotropic turbulence. *Science Bulletin, 61*(22), 1756–1765.
14. Ugarte, O. J., Bychkov, V., Sadek, J., et al. (2016). Critical role of blockage ratio for flame acceleration in channels with tightly spaced obstacles. *Physics of Fluids, 28*(9), 823–854.
15. Wang, C., Dong, X., Cao, J., et al. (2015). Experimental investigation of flame acceleration and deflagration-to-detonation transition characteristics using coal gas and air mixture. *Combustion Science and Technology, 187*(10–12), 1805–1820.

Chapter 6
Flame Quenching Characteristics of Gas Explosion in the Confined Spaces

6.1 Effect of Flame Arrester

6.1.1 Experimental Apparatus and Methods

The schematic diagram of the experimental apparatus was shown in Fig. 6.1. According to the ISO 16852 standard, the experimental apparatus was built, including an unprotected pipe ($L_1 = 2$ m), the flame arrester (flame arrester element and expansion chamber structure), a protected pipe ($L_2 = 2$ m), a high-voltage ignition system, a gas distribution system, a pressure acquisition system, a flame detection system, a data acquisition system, and a synchronous controller.

The unprotected and protected pipes were made of cast iron (DN65) with a design pressure of 15 MPa. The flame arrester was made of 304 stainless steel. Figure 6.2 showed the flame arrester with three porosities ($\alpha = 0.35$, 0.50, 0.65) and three element thicknesses (H = 60, 80, 100 mm). In order to effectively reduce the flame intensity acting on the flame arrester and analyze the effect of different expansion chambers, three types of expansion chamber structures (the extended type expansion chamber, the baffle type expansion chamber, and the conventional type expansion chamber) were used to connect to the flame arrester element, as shown in Fig. 6.3. The three expansion chamber structures have the same volume, but differ only in their internal structure. A ring-type baffle with an inner diameter of 70 mm was installed inside the baffle-type expansion chamber structure, and the distances of the baffle from the flame arrester element and flange end face were 50 mm and 215 mm respectively. The extended type expansion chamber structure was equipped with 155 mm pipe, and a 12 mm diameter round hole evenly distributed around it.

The high-voltage ignition system consisted of a high-temperature resistant ignition electrode and a KTD-A adjustable high energy igniter with an output ignition energy range from 1 to 20 J. The RCS2000-B high-precision gas distribution equipment was used for premixed gas preparation, and the gas distribution accuracy was 1%. The CY400 high frequency pressure transducers (accuracy: 0.25% FS; range: 0–1.5 MPa)

Z. Wang and X. Cao, *Gas Explosion and Its Protection Technology in Process Industries*, https://doi.org/10.1007/978-981-96-3121-6_6

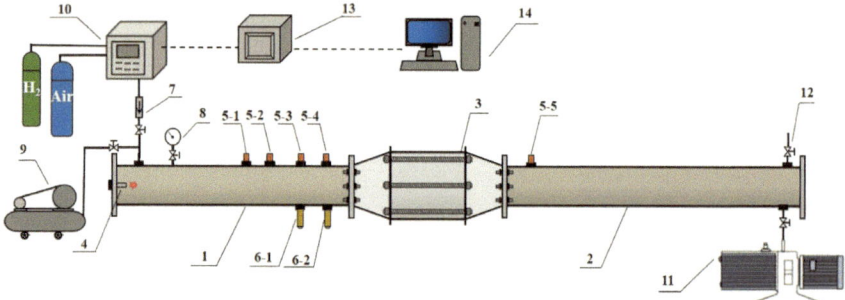

1-Unprotected pipe; 2-Protected pipe; 3-Flame arrester; 4-Ignition electrodes; 5-Flame detectors;
6-Pressure transducers; 7-Check valve; 8-Pressure gauge; 9-Air compressor; 10-Gas distribution system;
11-Vacuum pump; 12-Exhaust valve; 13-Synchronous controller; 14-Program control and acquisition system

Fig. 6.1 Schematic diagram of experimental apparatus

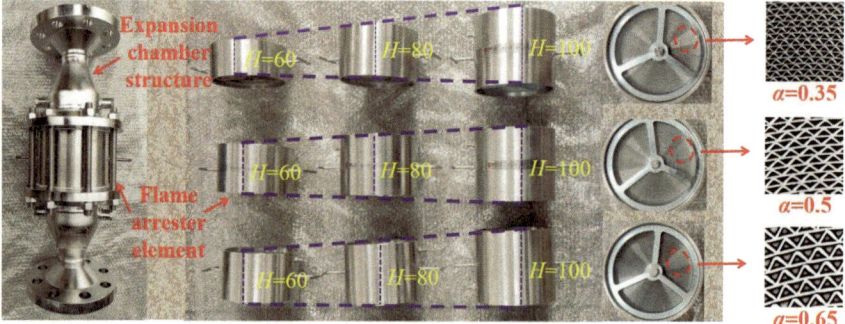

Fig. 6.2 Flame arresters with different porosities and element thicknesses

were used and installed on the unprotected pipe 400 mm and 200 mm from the left
end of the flame arrester, respectively. The CKG100 flame detectors were used and
installed in the unprotected and protected pipes respectively. The No. 1–4 flame
detectors were installed on the unprotected pipe, and the distances from the entrance
of the flame arrester were 1100 mm, 900 mm, 400 mm, and 200 mm respectively.
The No. 5 flame detector was installed on the protected pipe 200 mm from the right
end of the flame arrester to detect the success or failure of the flame quenching. The
TST6300 data acquisition instrument and synchronous control system were used to
realize the program control and data acquisition.

The experiments were carried out with the hydrogen-air premixed gas. The initial
pressure (P_0), hydrogen concentration (C_0) and ignition energy (E_0) could be adjusted
by the gas distribution system and high-voltage ignition system. The research
schemes for the effects of flame arrester structure parameters and initial conditions on
flame quenching performance were shown in Tables 6.1 and 6.2 respectively. Then
high-voltage discharge ignition and data acquisition were controlled cooperatively

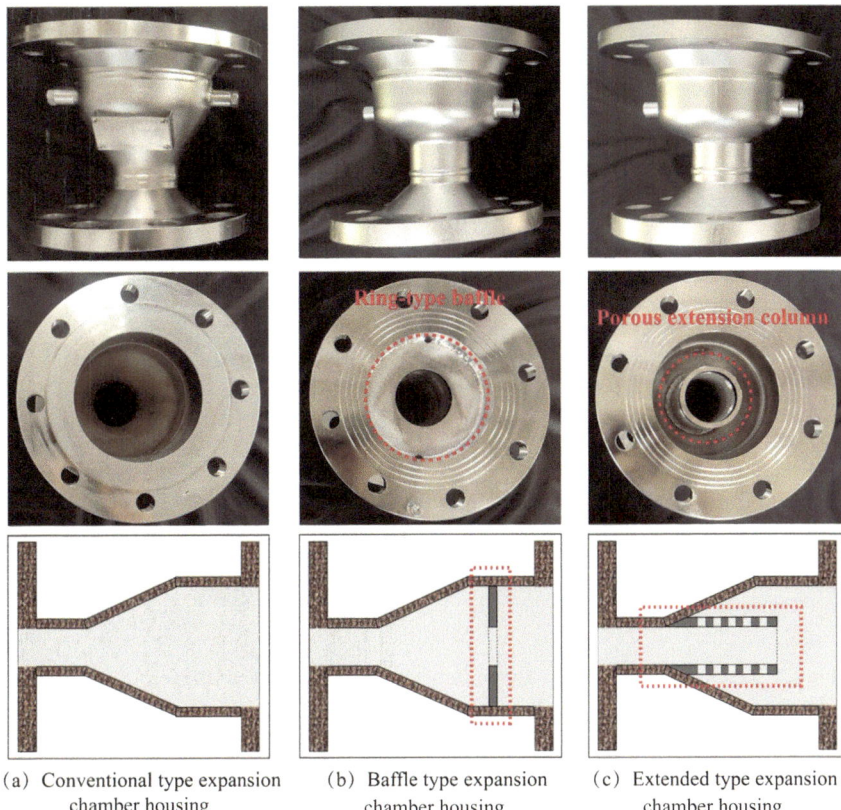

(a) Conventional type expansion chamber housing

(b) Baffle type expansion chamber housing

(c) Extended type expansion chamber housing

Fig. 6.3 Three types of expansion chamber structures

by the data acquisition system and the synchronous controller. To achieve reliable results, each experiment was repeated five times at least in this research. According to the measured results, the maximum deviations of the maximum overpressure and flame propagation velocity were 5.4% and 7.9% respectively, and the error bars have been added to the relevant figures.

In addition, based on the experimental results of single-factor effects on flame quenching performance, the effects of multi-factor interactions on the flame propagation velocity and overpressure entering the expansion chamber of the flame arrester were investigated using RSM. In the multi-factor interactions analysis of initial conditions, the Box–Behnken experiment design scheme (BBD) with 3-factor 3-level was used [1], and the experimental scheme was shown in Table 6.3. In the multi-factor interaction analysis of structure parameters, the Central Composite experiment design scheme (CCD) with 2-factor 3-level was used [2], and the experimental scheme was shown in Table 6.4. In addition, the multi-factor prediction models of the critical flame propagation velocity and overpressure under different structure

Table 6.1 Research scheme for the effect of initial conditions on flame quenching performance

No.	Influencing factors		Other initial conditions	Structure parameter values of flame arrester
A1	C_0	10%	$P_0 = 0$ MPa $E_0 = 13$ J	$\alpha = 0.35$ $H = 80$ mm
A2		20%		
A3		30%		
A4		40%		
A5		50%		
B1	P_0	− 0.05 MPa	$C_0 = 30\%$ $E_0 = 13$ J	
B2		− 0.025 MPa		
B3		0 MPa		
B4		0.025 MPa		
B5		0.05 MPa		
C1	C_0	3 J	$C_0 = 30\%$ $P_0 = 0$ MPa	
C2		8 J		
C3		13 J		
C4		18 J		

Table 6.2 Research scheme for the effect of flame arrester structure parameters on flame quenching performance

No.	Influencing factors		Other initial conditions	Structure parameter values of flame arrester
D1	α	0.35	$H = 100$ mm	$C_0 = 30\%$ $P_0 = 0$ MPa $E_0 = 13$ J
D2		0.5		
D3		0.65		
E1	H	60 mm	$\alpha = 0.35$	
E2		80 mm		
E3		100 mm		

parameters were established respectively, and the critical flame quenching criterion of flame arrester was proposed.

6.1.2 Effects of Initial Conditions and Structure Parameters

6.1.2.1 Initial Conditions

Figures 6.4 and 6.5 show the flame signal and overpressure curves measured at different positions respectively. As can be seen from Fig. 6.4, when the explosion flame passes the flame detector position, the flame detector detects the flame, which

Table 6.3 The BBD experiment design of initial conditions

No.	C_0 (%)	P_0 (MPa)	E_0 (J)
1	20	− 0.05 to 0.05	13
2	40	− 0.05 to 0.05	13
3	20	0.05	13
4	40	0.05	13
5	20	0	8
6	40	0	8
7	20	0	18
8	40	0	18
9	30	− 0.05 to 0.05	8
10	30	0.05	8
11	30	− 0.05 to 0.05	18
12	30	0.05	18
13	30	0	13
14	30	0	13
15	30	0	13
16	30	0	13
17	30	0	13

Table 6.4 The CCD experiment design of structure parameters

No.	α	H (mm)
1	0.35	60
2	0.65	60
3	0.35	100
4	0.65	100
5	0.35	80
6	0.65	80
7	0.5	60
8	0.5	100
9	0.5	80
10	0.5	80
11	0.5	80
12	0.5	80
13	0.5	80

causes the electrical signal curve to rise sharply. The flame propagation velocity was calculated from the distance of adjacent flame detectors and the flame signal interval time. By comparing the flame propagation velocity in different sections, it can be seen that the flame propagation velocity at the front end of the flame arrester increases

Fig. 6.4 The flame signals collected by the flame detectors

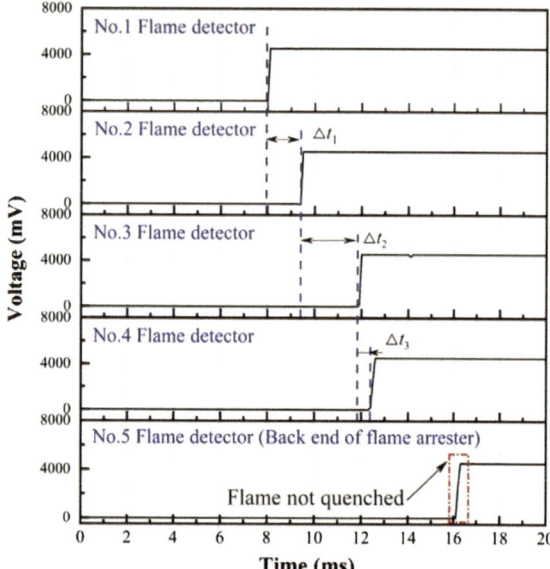

with the distance between the detectors and the flame arrester decreasing. Meanwhile, the flame propagation velocity was significantly reduced after the explosion flame enters the flame arrester compared the maximum flame propagation velocity before and after the flame arrester, which indicates that the flame arrester can effectively reduce the flame propagation velocity entering the flame arrester.

Figure 6.5 shows the overpressure curves measured at different positions near the front of the flame arrester. It can be seen that the process of the rising pressure curve at the front end of the flame arrester can be divided into four stages. In the I stage, the overpressure has no obvious change, because the flame propagation ability was weak, and the affected area was small in the initial stage of the explosion reaction. The remote pressure transducers do not feel the pressure wave, so the pressure curves in the I stage do not change. In the II stage, the explosion reaction rate increases with the explosion flame propagating along the pipe, and the pressure curve under the action of pressure wave shows a small upward change. With the accelerated propagation of the flame inside the pipe, the overpressure increases significantly. When the flame pass through the pressure transducers, the overpressure curves collected by the two transducers show a sharp rise in turn, and the maximum overpressure values were 0.46 MPa and 0.64 MPa respectively, which indicates that the overpressure curve measured by the transducer close to the flame arrester rises more obviously (III stage). When the explosion flame enters the expansion chamber and contacts with the flame arrester element, part of the flame and pressure wave could pass through the narrow channels. And the part of the pressure wave will propagate backward under the action of the flame arrester element, which leads to a secondary rise in the pressure curve, and the peak pressure appears at the same time (IV stage). However, because only part

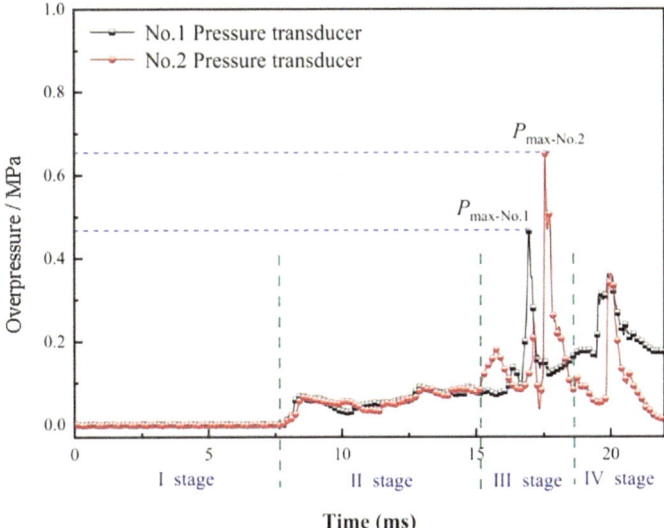

Fig. 6.5 The overpressure curves collected by pressure transducers

of the pressure wave passes through the flame arrester, the secondary pressure peak value was significantly reduced [3]. The overpressure value was evaluated by acoustic theory and the maximum overpressure value of 0.30 MPa was calculated based on the acoustic theory [4, 5]. However, the predicted value was less than the experimentally measured value ($P_{max\text{-}No.\ 1} = 0.46$ MPa) in the pipe. And the experimental value was close to the Sun's experimental value in confined space [6]. This was because the explosion flame propagates along the pipe due to the restraint of wall surface compared to the gas explosion in the open space [7]. The flame propagation was accelerated, which leads to the increase of the overpressure in the pipe. By comparing Figs. 6.4 and 6.5, it can be found that the maximum flame propagation velocity and overpressure value do not reach the detonation value of hydrogen-air premixed gas ($V_{CJ} = 1970$ m/s, $P_{CJ} = 1.54$ MPa) [8], so the experiment was the process of quenching the hydrogen deflagration flame by the crimped-ribbon pipe flame arrester. In addition, the No. 4 flame detector and No. 2 pressure transducer were located in the area entering the expansion chamber, so the maximum flame propagation velocity and overpressure measured at this location were regarded as the flame propagation velocity (V_f) and overpressure (P_f) entering the flame arrester respectively.

Figure 6.6a–c show the V_f, P_f, and the corresponding flame quenching results under different initial pressure (P_0), hydrogen concentration (C_0), and ignition energy (E_0) respectively. As shown in Fig. 6.6a, the V_f and P_f show a variation trend of first increasing and then decreasing with increasing the C_0. It can be observed that when the C_0 was 30%, which was close to the chemical equivalent concentration of hydrogen explosion, and the explosion intensity was the largest. The V_f and P_f reached the maximum values of 248 m/s and 0.45 MPa respectively, and the flame

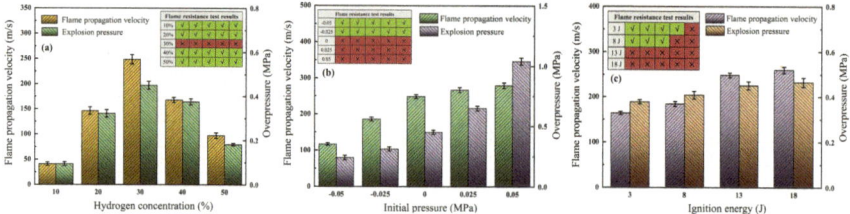

Fig. 6.6 The V_f, P_f, and flame quenching results under different initial conditions ($\checkmark$—no flame transmission, $\times$—flame transmission)

quenching failure at this time. The V_f and P_f values decrease with the C_0 moving away from 30%, and the explosion flame was successfully quenched. Based on the experimental results, the critical Peclet numbers of crimped ribbon flame arrester were calculated and evaluated for different equivalence ratios [9]. It can be found that the critical Peclet number was around 14, which was close to the Peclet number (21–28) calculated according to the MESG values in ISO 16852 [10]. This indicates that the flame arrester has better flame quenching performance for hydrogen explosion flame, which was consistent with the experimental results.

Meanwhile, it can be seen from Fig. 6.6b that both V_f and P_f increased significantly with increasing the P_0. Especially when the P_0 was 0.05 MPa, the V_f and P_f values reach 280 m/s and 1.04 MPa respectively. According to the experimental results, when the P_0 was $-$ 0.05 and $-$ 0.025 MPa, the explosion flame was effectively quenched. And the flame arrester failed when the P_0 was positive. This was because the hydrogen content per unit volume increases with increasing the P_0. The higher the chemical energy released per unit volume of hydrogen molecules when the premixed gas was ignited, and the hydrogen explosion intensity increased [11]. Therefore, the V_f and P_f values of hydrogen explosion increase with increasing the P_0, and the probability of flame quenching failure increases. In addition, the V_f and P_f also showed an increasing trend with increasing the E_0. Especially when the E_0 was 18 J, the V_f and P_f reach 259 m/s and 0.47 MPa respectively, as shown in Fig. 6.6c. According to the experimental results, the probability of successful flame quenching increases gradually when the E_0 was less than 13 J, and the flame arrester failed when the E_0 was greater than 13 J. This was because the energy to excite the chain reaction increases and the number of radicals involved in the reaction increases with increasing the E_0, resulting in an increase in the explosion reaction rate and the explosion intensity, so the explosion flame was difficult to be quenched. As can be seen above, the flame propagation velocity and overpressure entering the flame arrester were important factors that affect the success of the flame arrester. However, the initial conditions indirectly affect the flame quenching results by directly influencing the flame propagation velocity and overpressure entering the flame arrester.

6.1.2.2 Structure Parameters

Figure 6.7a, b show the V_f, P_f and the corresponding flame quenching results under different porosities (α) and element thicknesses (H) respectively. As shown in Fig. 6.7a, it can be observed that the V_f and P_f values increase progressively with increasing the α, this result is consistent with the experimental result of Jin [12]. The reflected pressure wave acting with the flame surface is significantly enhanced with decreasing the porosity, and the increase in flame turbulence makes the V_f and P_f values increased [13]. Meanwhile, the probability of flame quenching success increases significantly with decreasing the α. The flame arrester fails when the α is greater than 0.5, and the flame quenching success occurs when the α is 0.35. In addition, the V_f value decreases progressively (from 273 to 211 m/s) with increasing the H value, but the P_f value increases gradually (from 0.37 to 0.65 MPa), as shown in Fig. 6.7b. This is because relatively high flow resistance is produced by increasing the H, resulting in the higher pressure and lower flame propagation velocity in the front end of the flame arrester [14]. The flame arrester fails when the H is < 80 mm, and the flame quenching success of flame arrester occurs when the H is 100 mm, which indicates that the increase in the H can improve the flame quenching performance. With the H increasing, although the P_f value increases gradually, the V_f value decreases, resulting in the phenomenon of successful flame quenching by flame arrester. In summary, it can be seen that the structure parameters of flame arrester (the porosity and the element thickness) are related to the flame quenching performance of flame arrester. The changes in the structure parameters can affect the flame propagation velocity and overpressure entering the flame arrester, and could also affect the flame propagation process in the narrow channels of flame arrester, which in turn affects the flame quenching results.

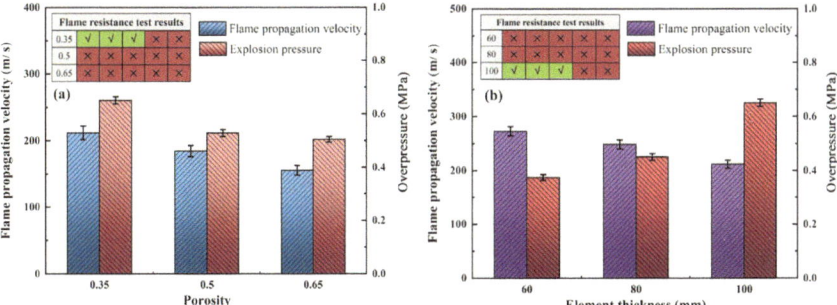

Fig. 6.7 The V_f, P_f, and flame quenching results under different structure parameters (√—no flame transmission, ×—flame transmission)

6.1.3 Interaction Effects of Multiple Factors

6.1.3.1 Interactions of Initial Conditions

Based on single-factor influence analysis, the interaction effects of initial conditions on the flame quenching performance are studied by RSM. The Box-Behnken experiment design scheme (BBD) with 3-factor 3-level is used. The response parameters are the flame propagation velocity (V_f) and the overpressure (P_f) entering the flame arrester, and the design parameters are the initial pressure (P_0), hydrogen concentration (C_0), and ignition energy (E_0) respectively. Meanwhile, the multi-factor prediction models of V_f and P_f under different initial conditions are obtained by RSM as follows:

$$
\begin{aligned}
V_f = & - 331.62732 + 31.55253^*C_0 + 1417.37250^*P_0 \\
& + 11.00443^*E_0 + 13.44730^*C_0P_0 + 0.054585^*C_0E_0 \\
& - 13.11200^*P_0E_0 - 0.517358^*C_0^2 - 15440.22000^*P_0^2 \\
& - 0.401310^*E_0^2
\end{aligned}
$$
$$R^2 = 0.9984 \tag{6.1}$$

$$
\begin{aligned}
P_f = & - 0.717351 + 0.070455^*C_0 + 7.57750^*P_0 + 0.006029^*E_0 \\
& - 0.028000^*C_0P_0 - 0.000190^*C_0E_0 + 0.055000^*P_0E_0 \\
& - 0.001085^*C_0^2 + 69.91000^*P_0^2 + 0.000171^*E_0^2
\end{aligned}
$$
$$R^2 = 0.9989 \tag{6.2}$$

To verify the validity of the prediction models, and determine the effect of the influencing factors and their interactions on the response parameters, the analysis of variance (ANOVA) is used to analyze the error of the models and the effect of each factor [15]. Tables 6.5 and 6.6 show the ANOVA results of prediction models of the V_f and P_f respectively. It can be found that the error probability (P-value) of the two prediction models is less than 0.0001, the models are "significant", and the Lack of Fit of both models is "not significant" [16]. Meanwhile, the R^2 values of the models are both greater than 0.90. In addition, as can be seen in Fig. 6.8, the predicted values of the models are in good agreement with experimental data. In summary, it is shown that the prediction models of V_f and P_f have good accuracy [17]. The smaller the P-value the greater the effect of the influencing factor on the response parameters [18]. It can be found from ANOVA that the P-values of P_0, C_0, and E_0 are less than 0.05, which indicates that all three influencing factors have a significant effect on V_f and P_f. Since the P-values of all three influencing factors are less than 0.0001, and the larger the F-value, the greater the effect of the design parameter on the response parameters, so the effect extent is analyzed by comparing the F-values corresponding to the design parameters. By comparing the F-value, it can be obtained that the effect order of the three factors in both models is: $P_0 > C_0 > E_0$.

In addition, it can be discovered from ANOVA that the interaction effects of initial conditions also have a significant effect on the prediction models. Figure 6.9a

Table 6.5 ANOVA for the response surface regression model of the V_f

Source	Sum of squares	df	Mean square	F-value	P-value	
Model	76,332.09	9	8481.34	1206.32	< 0.0001	Significant
C_0	1192.00	1	1192.00	169.54	< 0.0001	
P_0	54,472.15	1	54,472.15	7747.72	< 0.0001	
E_0	974.99	1	974.99	138.67	< 0.0001	
C_0P_0	180.83	1	180.83	25.72	0.0014	
C_0E_0	29.80	1	29.80	4.24	0.0785	
P_0E_0	42.98	1	42.98	6.11	0.0427	
C_0^2	11,269.87	1	11,269.87	1602.94	< 0.0001	
P_0^2	6273.69	1	6273.69	892.32	< 0.0001	
E_0^2	423.82	1	423.82	60.28	0.0001	
Residual	0.0004	7	0.0001			
Lack of fit	27.60	3	9.20	1.70	0.3033	Not significant
Pure error	21.61	4	5.40			
Cor total	76,381.30	16				

Table 6.6 ANOVA for the response surface regression model of the P_f

Source	Sum of squares	df	Mean square	F-value	P-value	
Model	1.29	9	0.1438	2308.95	< 0.0001	Significant
C_0	0.0067	1	0.0067	108.01	< 0.0001	
P_0	1.11	1	1.11	17,831.82	< 0.0001	
E_0	0.0046	1	0.0046	73.20	< 0.0001	
C_0P_0	0.0008	1	0.0008	12.59	0.0094	
C_0E_0	0.0004	1	0.0004	5.80	0.047	
P_0E_0	0.0008	1	0.0008	12.14	0.0102	
C_0^2	0.0495	1	0.0495	795.35	< 0.0001	
P_0^2	0.1286	1	0.1286	2064.70	< 0.0001	
E_0^2	0.0001	1	0.0001	1.24	0.3031	
Residual	0.0004	7	0.0001			
Lack of fit	0.0002	3	0.0001	0.879	0.5232	Not significant
Pure error	0.0003	4	0.0001			
Cor total	1.29	16				

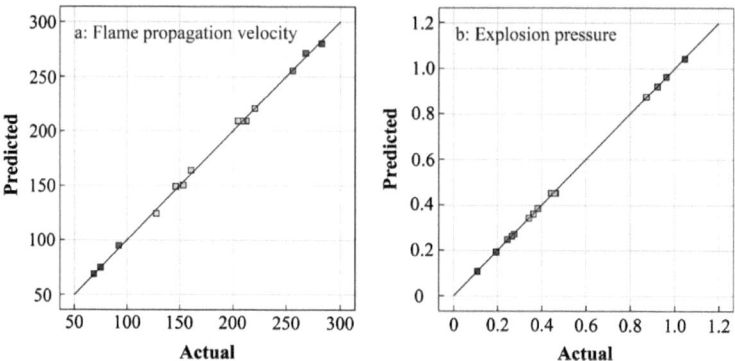

Fig. 6.8 The predicted versus actual values plot for V_f and P_f

shows the interaction effects of C_0 and P_0 on V_f and P_f respectively. It can be seen that when the C_0 is the equivalent value, both V_f and P_f increase significantly with increasing the P_0. When the C_0 is far from the equivalent value, both V_f and P_f show an increasing trend with increasing the P_0. Figure 6.9b shows the interaction effects of C_0 and E_0 on V_f and P_f respectively. With increasing the C_0 and E_0, both V_f and P_f show a trend of increasing first and then decreasing, and the C_0 has a greater effect on V_f and P_f than the E_0. Figure 6.9c shows the interaction effects of P_0 and E_0 on V_f and P_f respectively. The V_f and P_f values increase with increasing the E_0 and P_0. Compared with the E_0, the P_0 has a more significant effect on the V_f and P_f values. According to Tables 6.5 and 6.6, comparing the P-values of each two factors' interaction in the initial conditions, it is found that the influence order of the interaction effects on V_f and P_f is: (C_0 and P_0) > (P_0 and E_0) > (C_0 and E_0).

6.1.3.2 Interactions of Structure Parameters

Based on single-factor influence analysis, the interaction effects of structure parameters on the flame quenching performance were studied by RSM. The Central Composite experiment design scheme (CCD) with 2-factor 3-level is used, as shown in Table 6.4. The response parameters were the flame propagation velocity (V_f) and the overpressure (P_f) entering the flame arrester, and the design parameters were flame arrester porosity (α) and element thickness (H). Meanwhile, the multi-factor prediction models of V_f and P_f under different structure parameters were obtained by RSM as follows:

$$V_f = -617.06505 - 325.72632^*\alpha - 5.48886^*H - 3.09970^*\alpha H$$
$$+ 343.41176^*\alpha^2 + 0.035203^*H^2$$
$$R^2 = 0.9902 \tag{6.3}$$

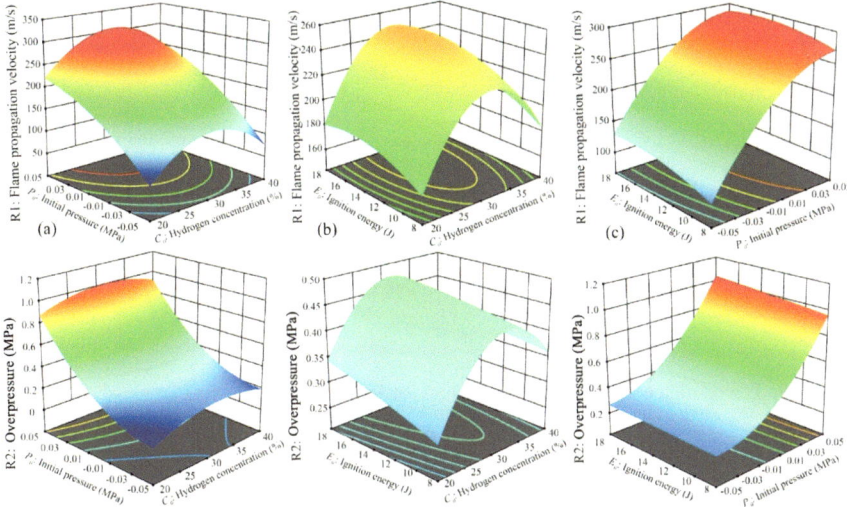

Fig. 6.9 The response surface plot for interaction effects of **a** C_0 and P_0, **b** C_0 and E_0, **c** P_0 and E_0 on the V_f and P_f

$$P_f = 0.697762 - 0.501303*\alpha - 0.005935*H - 0.007083*\alpha H$$
$$+ 0.727969*\alpha^2 + 0.000087*H^2$$
$$R^2 = 0.9921 \tag{6.4}$$

According to the P-values corresponding to the V_f and P_f prediction models in ANOVA, and the comparison results between the predicted and experimental values in Fig. 6.10, as well as the R^2 value of the equations, it can be found that the prediction models of V_f and P_f have good accuracy with the experimental values. The P-values of structure parameters are less than 0.05 according to the ANOVA results, which indicates that α and H have a significant effect on V_f and P_f. By comparing the F-values corresponding to the design parameters, it can be found that the effect of α (F-value = 377.46) on V_f model is greater than that of H (F-value = 250.20), while the effect of H (F-value = 588.64) on P_f. model is greater than that of α (F-value = 191.14). Figure 6.11 presents the response surface plots for the interaction effects of α and H on V_f and P_f. With the decrease of α and H, the V_f value rises significantly, and the α has a greater effect on V_f model than H, as shown in Fig. 6.11a. The P_f value increases significantly with increasing the H and the decreasing the α. The H has a greater effect on the P_f model than α, as shown in Fig. 6.11b.

In summary, the changes of the initial conditions would cause the changes of the V_f and P_f when the flame arrester structure parameters values are fixed. However, the critical flame propagation velocity (V_c) and overpressure (P_c) at which the flame arrester can successfully quench flame would not change with changing the initial conditions. For example, when the flame arrester porosity and element thickness are

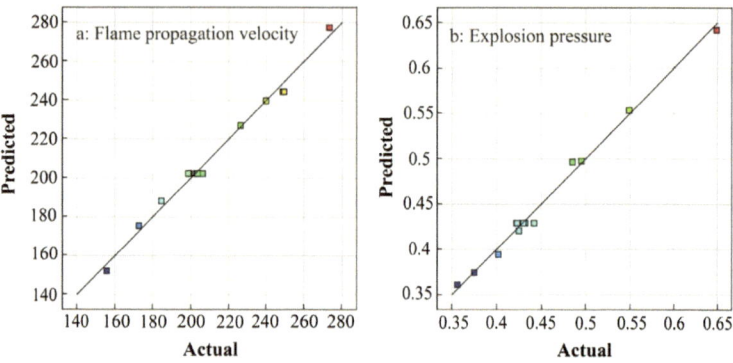

Fig. 6.10 The predicted versus actual values plot for V_f and P_f

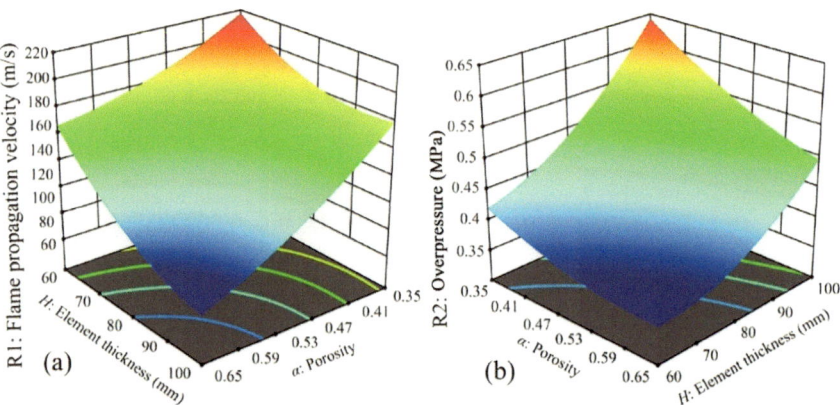

Fig. 6.11 The response surface plot for interaction effects of different structure parameters on the V_f and P_f

0.35 and 80 mm respectively, the V_c and P_c of the flame arrester under different initial conditions are 183 m/s and 0.51 MPa respectively. However, when the initial conditions values are fixed, the changes in the structure parameters would cause changes in the V_c and P_c values. For example, when the porosity (0.35–0.65) and element thickness (60–100 mm) of the flame arresters were changed, the V_c value increases from 65 to 184 m/s, and the P_c value increases from 0.15 to 0.61 MPa. Therefore, it can be concluded that the structure parameters of the flame arresters are important influencing factors for the V_c and P_c.

6.2   Effects of Obstacle Parameter on Explosion Resistance Performance of Flame Arrester

6.2.1   Experimental Apparatus and Methods

Figure 6.12 illustrates a schematic diagram of experimental apparatus for studying the explosion resistance performance of hydrogen crimped-ribbon flame arrester under the presence of obstacles. The main body included the explosion pipe (including an unprotected pipe (L1 = 2 m) and a protected pipe (L2 = 2 m)), a flame arrester (including flame arrester element and expansion chamber structure), an obstacle group, a gas distribution system, an ignition system, a flame acquisition system, a pressure acquisition system, a temperature acquisition system and a program control and data acquisition system.

The explosion pipe was made of cast iron (DN65) with a design pressure of 15 MPa. Two ends of pipe were sealed by Q345R flanges. The flame arrester element was selected with the thickness (H) was 100 mm, the ripple height (λ) was 0.2 mm and the porosity (α) was 0.35. To clarify the effect of obstacle parameters on the explosion resistance performance, a detachable obstacle group was used to change different working conditions. The blockage rate of obstacle was calculated based on the formula (BR = 1 − r2/R2), and their distance was changed by adjusting the location of adjacent obstacles (the distance between adjacent obstacles was equal).

The premixed gas was prepared by a RCS2000-B high-precision gas distribution instrument with 1% accuracy. The ignition system consisted of a high-temperature resistant electrode and an adjustable high-energy igniter (KTD-A). Four and one CKG100 flame detectors were installed on the unprotected pipe and the protected pipe to measure the flame propagation velocity. Two and one CY400 high-frequency pressure transducers were installed at the front and rear ends of flame arrester to measure the pressure. Two S-type high-frequency thermocouple were installed at the

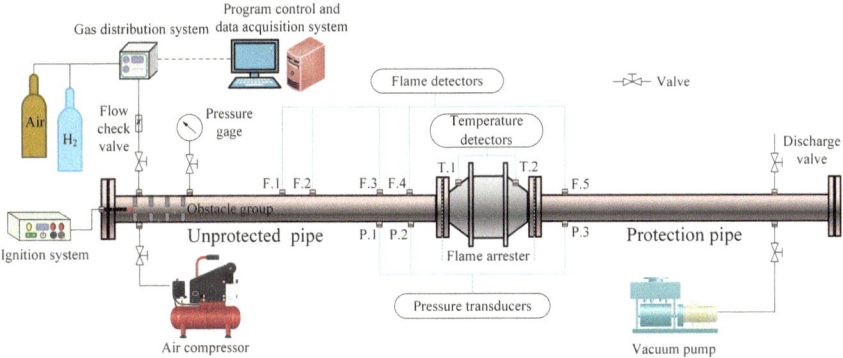

Fig. 6.12   Schematic diagram of experimental apparatus

Table 6.7 Installation distance between detectors

Location 1	Location 2	Distance (mm)
F.1 Flame detector	F.2 Flame detector	200
F.2 Flame detector	F.3 Flame detector	500
F.3 Flame detector	F.4 Flame detector	200
F.4 Flame detector	Left end flange of flame arrester	150
F.5 Flame detector	Right end flange of flame arrester	150
P.1 Pressure transducer	P.2 Pressure transducer	200
P.2 Pressure transducer	Left end flange of flame arrester	150
P.3 Pressure transducer	Right end flange of flame arrester	150
T.1 Temperature detector	Left end flange of flame arrester	200
T.1 Temperature detector	T.2 Temperature detector	230
T.2 Temperature detector	Right end flange of flame arrester	200

front and rear ends of flame arrester to measure the temperature. The specific installation location is shown in Table 6.7. Meanwhile, the acquisition system (TST6300) was controlled by program language to realize the output of program signal and data acquisition.

The premixed gas with 20% hydrogen concentration (Volume fraction) was selected. And the ignition energy was 9 J, and the initial ignition pressure was atmospheric pressure. Under the case without obstacle, the flame velocity, overpressure and flame temperature entering the front end of flame arrester were expressed by V_0, P_0 and T_0, respectively. As obstacles existed, the explosion parameter entering the flame arrester was expressed by V_f, P_f and T_f. And the explosion parameter at the rear end of flame arrester was expressed by V_r, P_r and T_r. The flame resistance result was expressed by R_s (S and F represent the flame resistance success and failure, respectively). Each working condition was repeated at least five times to obtain accurate and reliable experimental results. According to the results analysis, the maximum deviations of flame propagation velocity, overpressure and flame temperature were 8.6%, 6.9% and 2.2%, respectively, and the error bars have been added to the corresponding figures.

6.2.2 Obstacle Location

Figure 6.13 presents the flame propagation velocity at the front and rear ends of flame arrester under different obstacle locations ($N = 1$; $BR = 0.7$). It can be seen that the flame propagation velocity showed a trend of first increasing and then decreasing with the increase of obstacle location. The flame propagation velocity reached the maximum as L was 650 mm. In particular, the flame propagation velocity of front end was 5.20 times than that under the case without obstacle ($V_f = 5.2V_0$). The flame propagation inside the pipe was accelerated under the action of obstacle ($V_f >$ V_0). However, the flame acceleration was less affected by the obstacle as the obstacle was close to the ignition source ($L < 325$ mm) [19]. The flame quenching inside the flame arrester under this condition, and no obvious voltage signal was monitored by the flame detector of rear end. As the obstacle was far away from the ignition source (325 mm $< L <$ 845 mm), the flame propagation velocity was obviously increased and it could pass through the flame arrester under the combined action of self-acceleration and obstacle [20]. Thereafter, the flame acceleration effect began to decrease as the obstacle location continued to increase ($L > 845$ mm) [21, 22]. This further led to a decrease in the flame velocity entering the flame arrester. Based on the flame resistance results and the corresponding flame propagation velocity, the critical flame resistance velocity was determined ($V_c = 227.8$ m/s). Meanwhile, the flame velocity showed a decrease trend due to the blocking effect of flame arrester ($V_r < V_f$).

Figure 6.14 illustrates the overpressure near the front and rear ends of flame arrester under different obstacle locations ($N = 1$; $BR = 0.7$). The overpressure of front end showed a trend of first increasing and then decreasing with the increase of obstacle location. The pressure of front end reached the maximum as L was 650 mm ($P_f = 4.0P_0$). However, there was an obvious difference in the overpressure of rear end [23]. As the obstacle was close to the ignition source ($L < 325$ mm), the overpressure was slightly increased under the action of obstacle. The explosion shock wave intensity inside the flame arrester was obviously weakened under this condition, further resulting in the pressure history of rear end presented one peak. As the obstacle was far away from the ignition source (325 mm $< L <$ 845 mm),

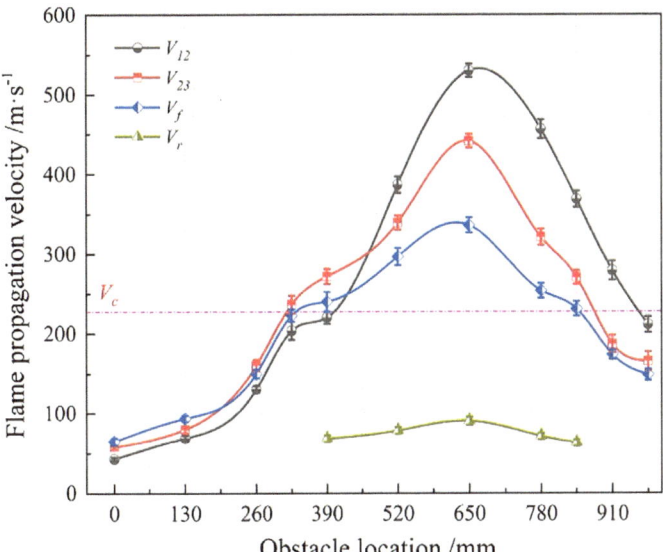

Fig. 6.13 The flame propagation velocity under different obstacle locations (N = 1; BR = 0.7)

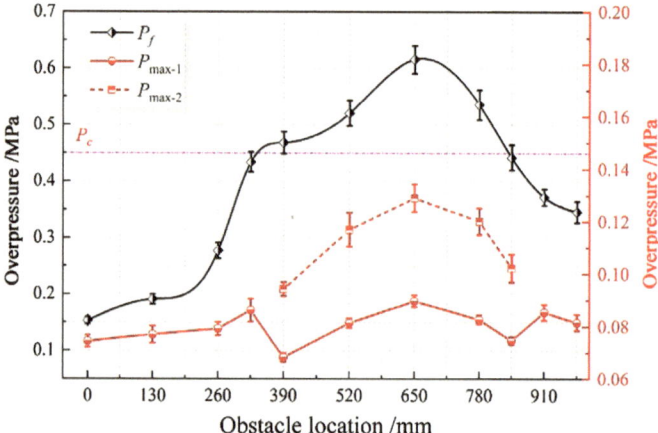

Fig. 6.14 The overpressure under different obstacle locations ($N = 1$; $BR = 0.7$)

the overpressure was obviously increased under the combined action of flame self-acceleration and obstacle. This led to the flame resistance failure and the pressure history of rear end appeared two peaks. Besides, the shock wave inside the narrow channel was greatly weakened during propagation, which led to a lower peak in the pressure history I when the flame resistance failed. The peak of pressure history II showed the same trend as the overpressure of front end during this range. After that, the promotion effect of obstacle on the shock wave was weakened as the obstacle location continued to increase ($L > 845$ mm). This led to a decrease in the overpressure entering the flame arrester and the flame resistance was successful. Meanwhile, the critical flame resistance pressure was determined ($P_c = 0.45$ MPa) based on the flame resistance results and the corresponding overpressure.

Figure 6.15 shows the flame temperature at the front and rear ends of flame arrester and the corresponding flame resistance results under different obstacle locations ($N = 1$; $BR = 0.7$). It can be seen that the flame temperature showed a trend of first increasing and then decreasing with the increase of obstacle location. As L was 650 mm, the flame temperature at the front and rear ends all reached the maximum and the T_f was greater than T_r. However, the flame flow field was less disturbed by obstacle as the obstacle was close to the ignition source ($L < 325$ mm). The flame temperature entering the flame arrester was low under this condition, which resulted in no obvious increase in the temperature of rear end. This indicated that the flame resistance was successful. As the obstacle was far away from the ignition source (325 mm $< L <$ 845 mm), the flame temperature was obviously increased under the combined action of self-acceleration and obstacle. In particular, the flame could pass through the flame arrester and ignite the premixed gas of rear end, further resulting in the flame resistance failure and an obvious increase in the temperature of rear end. The disturbance of flame flow field by the obstacle began to decrease as the obstacle location continued to increase ($L > 845$ mm), which led to the decrease of flame

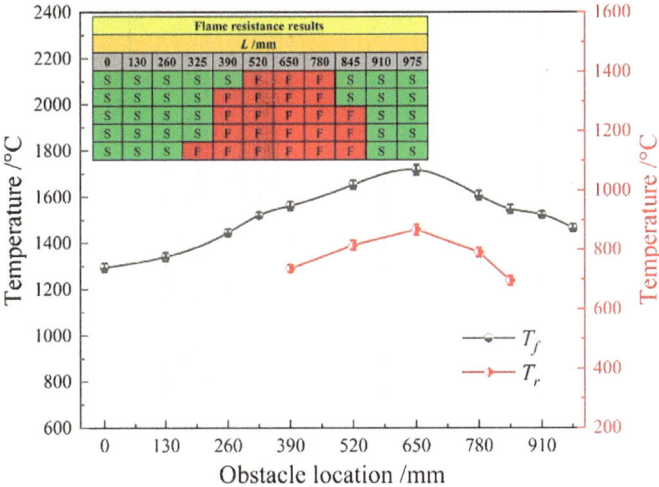

Fig. 6.15 The temperature and flame resistance results under different obstacle locations ($N = 1$; $BR = 0.7$)

temperature entering the flame arrester. In particular, the temperature history of rear end did not show an obvious increase trend.

According to the flame resistance results, the flame resistance failure began to occur as L was 325 mm, and the probability of flame resistance success showed a trend of first decreasing and then increasing with the increase of obstacle location. As 520 mm $\leq L \leq$ 780 mm, the flame resistance all failed. However, the phenomenon of flame resistance success was presented again as L was 845 mm. It can be seen that the probability of flame resistance success was related to the obstacle location and was greatly affected by it. Besides, based on explosion parameters and flame resistance results, the quantitative prediction models of V_f, P_f and T_f under different obstacle locations were proposed, as shown in (6.5), (6.6), (6.7).

$$V_f = 48.61414 + \left(174{,}806.42809 / \left(498.00896 \times \sqrt{\pi/2}\right)\right)$$
$$\times \exp\left(-2 \times \left(\frac{(L - 602.29066)}{498.00896}\right)^2\right)$$
$$R^2 = 0.9868 \tag{6.5}$$

$$P_f = 0.12109 + \left(315.63988 / \left(522.11371 \times \sqrt{\pi/2}\right)\right)$$
$$\times \exp\left(-2 \times ((L - 622.48957)/522.11371)^2\right)$$
$$R^2 = 0.9868 \tag{6.6}$$

$$T_f = 1247.02629 + \left(321{,}198.21616 / \left(581.08771 \times \sqrt{\pi/2}\right)\right)$$

$$\times \exp\left(-2 \times ((L - 622.74745)/581.08771)^2\right)$$
$$R^2 = 0.9843 \tag{6.7}$$

6.2.3 Obstacle Number

Figure 6.16a–c illustrate the effect of obstacle number on the explosion parameters and the corresponding flame resistance results under different blockage rates ($D = 120$ mm; $BR = 0.5, 0.6$ and 0.7).

The explosion parameters (V_f, P_f and T_f) showed an increase trend with the increase of obstacle number [24, 25], and the rising degree was continuously increased. In particular, the explosion parameters reached the maximum under the action of obstacle number was 8. The effect of blockage rate on the explosion parameter was not obvious as $N < 4$. However, the blockage rate had an obvious effect on the explosion parameter as $N \geq 4$. The above two obstacle parameters affected the probability of flame resistance success by affecting the explosion parameter. The flame resistance results were successful under three blockage rates as $N < 4$. The flame resistance results were failed under three blocking rates as $N > 4$. The probability of flame success resistance was increased with the increase of blockage rate as N was 4. This indicates that the obstacle number could significantly affect the explosion parameter entering the flame arrester, and then affect the flame resistance result. In particular, the flame resistance failure mostly occurred when the obstacle number was larger.

6.2.4 Obstacle Blockage Rate

Figure 6.17a–c show the effect of blockage rate on the explosion parameters and the corresponding flame resistance results under three obstacle numbers ($D = 120$ mm; $N = 4, 6$ and 8). The explosion parameters (V_f, P_f and T_f) near the front end of flame arrester showed a trend of first increasing and then decreasing with the increase of blockage rate [26, 27]. In particular, the explosion parameters reached the maximum under the action of obstacles as N was 8 and BR was 0.5 ($V_f = 11.8V_0$, $P_f = 9.8P_0$ and $T_f = 1.4T_0$). And the probability of flame success resistance was the lowest. Besides, the explosion parameter was continuously increased with the increase of obstacle number. Combined with the explosion parameters, the probability of flame success resistance showed a trend of first decreasing and then increasing with the increase of blockage rate as N was 4. However, the flame resistance all failed under five blockage rates as the obstacle number continued to increase ($N = 6$ and 8). This indicates that the blockage rate could significantly affect the explosion parameters entering the flame arrester, and then affect the probability of flame resistance success. In particular, the flame resistance failure was more likely to occur when the blockage rate was larger.

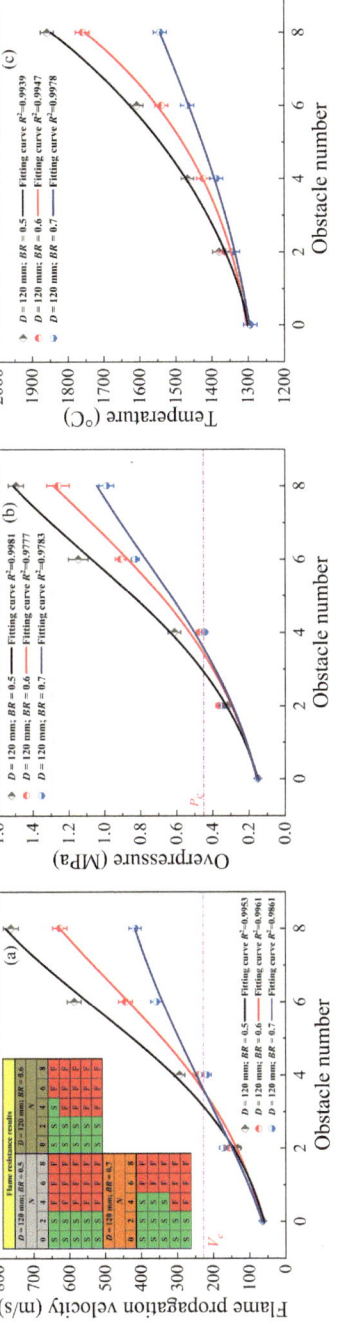

Fig. 6.16 Effects of obstacle number under different obstacle blockage rates ($D = 120$ mm; $BR = 0.5$, 0.6 and 0.7)

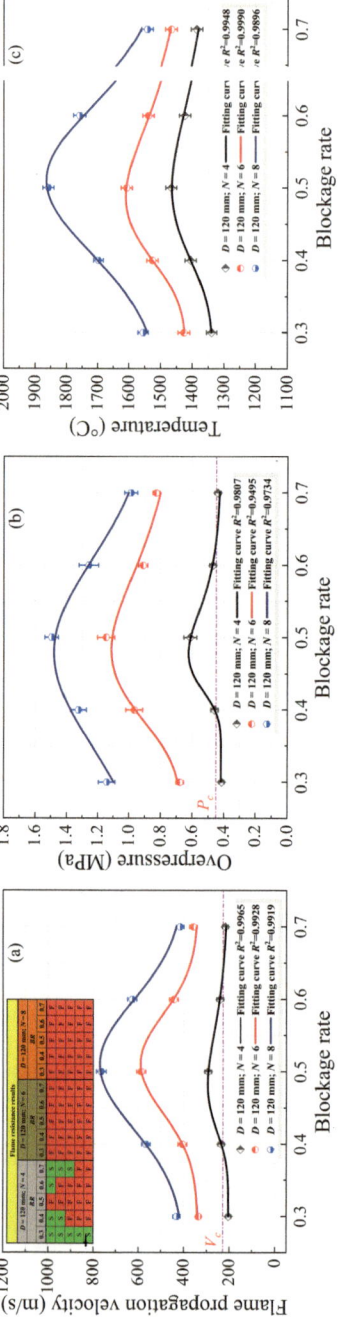

Fig. 6.17 Effects of blockage rate under different obstacle numbers ($D = 120$ mm; $N = 4$, 6 and 8)

References

1. Mohammed, S. S., Sreejith, N. K., Kumar, R. D., et al. (2021). Parametric optimization of hole geometry in laser drilled Inconel 825 by grey based response surface methodology. *Materials Today: Proceedings, 47*, 5410–5415.
2. Le, A. T., Tan, Z. H., Sivakumar, R., et al. (2021). Predicting the photocatalytic performance of metal/metal oxide coupled TiO_2 particles using Response Surface Methodology (RSM). *Materials Chemistry and Physics, 269*(13), 124739.
3. Wen, X., Xie, M., Yu, M., et al. (2013). Porous media quenching behaviors of gas deflagration in the presence of obstacles. *Experimental Thermal and Fluid Science, 50*, 37–44.
4. Kim, W. K., Mogi, T., et al. (2014). Effect of propagation behaviour of expanding spherical flames on the blast wave generated during unconfined gas explosions. *Fuel, 128*, 396–403.
5. Kim, W. K., Mogi, T., & Dobashi, R. (2013). Fundamental study on accidental explosion behavior of hydrogen-air mixtures in an open space. *International Journal of Hydrogen Energy, 38*(19), 8024–8029.
6. Sun, S., Liu, G., Liu, J., et al. (2017). Effect of porosity and element thickness on flame quenching for in-line crimped-ribbon flame arresters. *Journal of Loss Prevention in the Process Industries, 50*, 221–228.
7. Lu, Y., Wang, Z., Cao, X., et al. (2021). Interaction mechanism of wire mesh inhibition and ducted venting on methane explosion. *Fuel, 304*, 121343.
8. Werling, L., Lauck, F., Freudenmann, D., et al. (2017). Experimental investigation of the flame propagation and flashback behavior of a green propellant consisting of N_2O and C_2H_4. *Journal of Energy and Power Engineering, 11*(12), 735–752.
9. Thomas, G., Oakley, G., & Bambrey, R. (2020). Fundamental studies of explosion arrester mitigation mechanisms. *Process Safety and Environmental Protection, 137*, 15–33.
10. Zhang, K., Wang, Z., Gong, J., et al. (2017). Experimental study of effects of ignition position, initial pressure and pipe length on H_2-air explosion in linked vessels. *Journal of Loss Prevention in the Process Industries, 50*, 295–300.
11. Jin, K., Duan, Q., Chen, J., et al. (2017). Experimental study on the influence of multi-layer wire mesh on dynamics of premixed hydrogen-air flame propagation in a closed duct. *International Journal of Hydrogen Energy, 42*(21), 14809–14820.
12. Yue, J., Long, W., Liu, H., et al. (2021). A novel detonation arrester containing a large disk with long triangular slits: Design and numerical simulation. *Process Safety Progress, 40*(1), 12176.
13. Sun, S. C., Shu, Y., Feng, Y., et al. (2018). Numerical simulation of detonation wave propagation and quenching process in in-line crimped-ribbon flame arrester. *Cogent Engineering, 5*(1), 1–21.
14. Bao, L., Wang, P., Dang, W., et al. (2021). Experimental study on detonation flame penetrating through flame arrester. *Journal of Loss Prevention in the Process Industries, 72*, 104529.
15. Khoobbakht, G., Najafi, G., & Karimi, M. (2016). Optimization of operating factors and blended levels of diesel, biodiesel and ethanol fuels to minimize exhaust emissions of diesel engine using response surface methodology. *Applied Thermal Engineering, 99*, 1006–1017.
16. Wasserstein, R. L., & Lazar, N. A. (2016). The ASA statement on p-values: Context, process, and purpose. *American Statistician, 70*(2), 129–133.
17. Prasad, G. A., Murugan, P. C., Wincy, W. B., et al. (2021). Response surface methodology to predict the performance and emission characteristics of gas-diesel engine working on producer gases of non-uniform calorific values. *Energy, 234*, 121225.
18. Pereira, L. M. S., Milan, T. M., & Tapia-Blácido, D. R. (2021). Using response surface methodology (RSM) to optimize 2G bioethanol production: A review. *Biomass and Bioenergy, 151*, 106166.
19. Zheng, K., Song, C., Yang, X., et al. (2022). Effect of obstacle location on explosion dynamics of premixed H_2/CO/air mixtures in a closed duct. *Fuel, 324*, 124703.

20. Han, S., Yu, M., Yang, X., et al. (2020). Effects of obstacle position and hydrogen volume fraction on premixed syngas-air flame acceleration. *International Journal of Hydrogen Energy, 45*(53), 29518–29532.
21. Qiming, X., Guohua, C., Qiang, Z., et al. (2022). Numerical simulation study and dimensional analysis of hydrogen explosion characteristics in a closed rectangular duct with obstacles. *International Journal of Hydrogen Energy, 47*(92), 39288–39301.
22. Wu, Q., Yu, M., & Zheng, K. (2022). Experimental investigation on the effect of obstacle position on the explosion behaviors of the non-uniform methane/air mixture. *Fuel, 320*, 123989.
23. Lv, X., Zheng, L., Zhang, Y., et al. (2016). Combined effects of obstacle position and equivalence ratio on overpressure of premixed hydrogen–air explosion. *International Journal of Hydrogen Energy, 41*(39), 17740–17749.
24. Elshimy, M., Ibrahim, S., & Malalasekera, W. (2020). Numerical studies of premixed hydrogen/air flames in a small-scale combustion chamber with varied area blockage ratio. *International Journal of Hydrogen Energy, 45*(29), 14979–14990.
25. Qin, Y., & Chen, X. (2021). Flame propagation of premixed hydrogen-air explosion in a closed duct with obstacles. *International Journal of Hydrogen Energy, 46*(2), 2684–2701.
26. Goodwin, G. B., Houim, R. W., & Oran, E. S. (2016). Effect of decreasing blockage ratio on DDT in small channels with obstacles. *Combustion and Flame, 173*, 16–26.
27. Xiao, H., & Oran, E. S. (2020). Flame acceleration and deflagration-to-detonation transition in hydrogen-air mixture in a channel with an array of obstacles of different shapes. *Combustion and Flame, 220*, 378–393.

Chapter 7
Resistance/Venting Characteristics of Gas Explosion in the Confined Spaces

7.1 Resistance/Venting Characteristics of Gas Explosion in the Confined Spaces

7.1.1 Experimental Apparatus and Methods

7.1.1.1 Experimental Apparatus

The experimental apparatus is illustrated in Fig. 7.1. The spherical vessel is 600 mm in diameter and 110 L in volume. Moreover, there is a duct connected to the vessel, which is 250 mm in length. The two cylindrical pipes are 2000 mm in length and 60 mm in internal diameter are connected by flanges, then connect the pipe to the duct. Finally, the end of the pipe is sealed with a blind flange, and a closed pipe-connected spherical vessel is formed. A spark-plug 1 and a charge/exhaust valve 2 are installed on the spherical vessel. The pressure transmitters are flushed to the surface of the spherical vessel, both sides of the porous material and the end of pipes; the positions of the pressure transmitters are denoted as 4, 5, 6 and 7.

7.1.1.2 Experimental Methods

In this work, the porous material was fixed on a customized thin iron hoop with 0.5 mm stainless steel wire in the process of experiment, the sketch of the porous material fixed in the pipe is shown in Fig. 7.2. Three categories of the porous materials, foam Fe–Ni, foam ceramic Al_2O_3, and foam ceramic SiC were selected as objects for the CH_4/Air explosion suppression experiment. The geometric parameters of the porous materials are shown in Table 7.1.

The initial pressure of the experiment was set to 0 MPa (the pressures in the paper are gauge pressure). The closed pipe-connected spherical vessel was evacuated to

Z. Wang and X. Cao, *Gas Explosion and Its Protection Technology in Process Industries*, https://doi.org/10.1007/978-981-96-3121-6_7

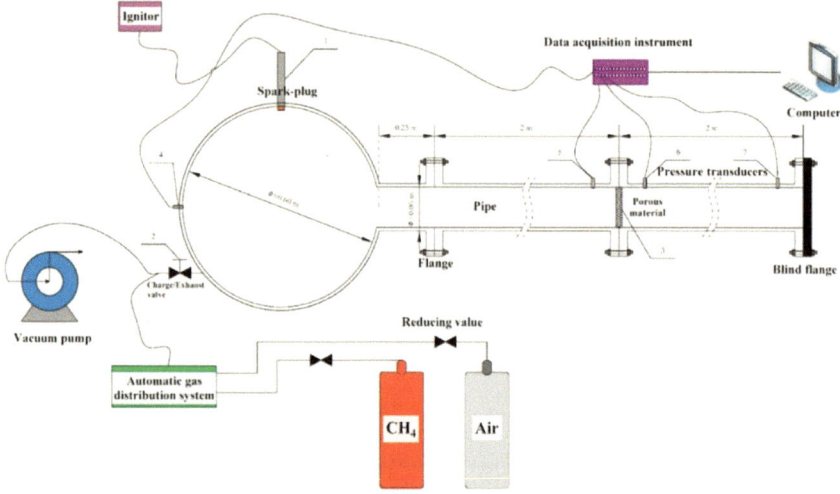

Fig. 7.1 Structural schematic of the experimental apparatus; 1: spark-plug; 2: charge/exhaust valve; 3: porous material; 4, 5, 6, 7: pressure transmitters

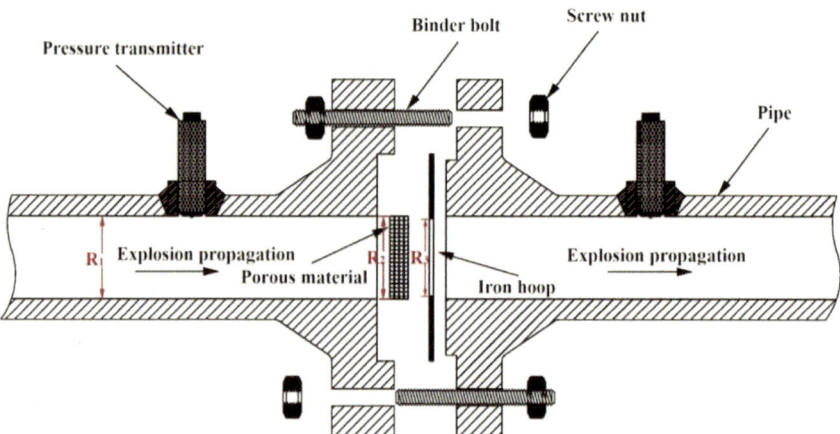

Fig. 7.2 Schematic drawing of fixing mode of porous materials

− 0.095 MPa by the vacuum pump. Then the CH_4/Air mixture at a concentration of 10% was transferred into the apparatus by an RCS2000-B automatic gas distribution system, until the gauge pressure indicated 0 MPa. Before ignition, the gas mixture was allowed to mix for 3–5 min to ensure the quiescence and homogeneity. A high-tension spark plug located on the upper wall of the spherical vessel was activated to ignite the flammable premixed gases, and the ignition energy was 6 J. The HM90-H3-2 high-frequency pressure transmitters were used to measure the overpressure. The JV5231 multichannel data acquisition instrument was adopted to collect synchronous

Table 7.1 Geometrical parameters of porous materials

No.	Porous material	Thickness (mm)	Pore size (PPI)	Volume density (g/cm³)	Open-cell rate (%)
1	Fe–Ni	10	90	0.4172	≥ 98
		10	40	0.2694	≥ 98
2	SiC	20	20	0.6030	80–90
		20	10	0.5795	80–90
3	Al₂O₃	10	50	0.5803	80–90
		10	30	0.7249	80–90

Note PPI means the number of pores per inch

data. After a test, the experimental apparatus was flushed by air three times to avoid the effect of residual exhaust gases. Each group of the experiment was repeated at least three times to confirm the reproducibility.

7.1.1.3 Experimental Conditions

The aim of this work it to investigate the suppression effect of porous materials on CH_4/Air mixtures explosion. Three categories of porous materials were used in the experimental study. The detailed experimental scheme is shown in Table 7.2.

7.1.2 Explosion Characteristics Without Porous Materials

As shown in Table 7.2 (i.e., Group A), a blank experiment was carried out before conducting the study on the explosion suppression of the porous material in the closed pipe-connected spherical vessel. The blank experiment consisted of two sets of tests, one of which was without any obstruction, i.e., A-1; while the other included a fixed thin iron hoop (i.e., A-2) at the location (i.e., position 3) where the porous material was placed. Generally, there exist inherent oscillations in the explosion evolution curve, the sources of the oscillations include combustion noise, mechanical vibration and so

Table 7.2 Experimental scheme of porous materials for explosion suppression

Group	Experimental scheme	
A	(1) None	(2) Iron hoop
B	(1) Fe–Ni 10 mm/90 PPI	(2) Fe–Ni 10 mm/40 PPI
	(3) SiC 20 mm/20 PPI	(4) SiC 20 mm/10 PPI
	(5) Al₂O₃ 10 mm/50 PPI	(6) Al₂O₃ 10 mm/30 PPI

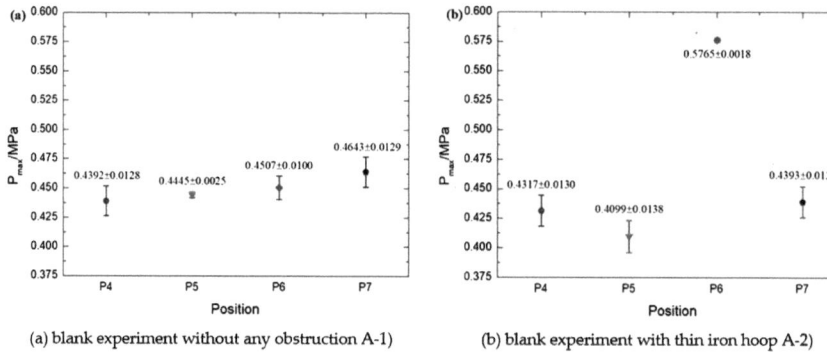

(a) blank experiment without any obstruction A-1)   (b) blank experiment with thin iron hoop A-2)

Fig. 7.3 Maximum overpressure and standard deviation of the blank experiment

on [1]. Thus, the explosion evolution curve is smoothed by the method of Savitzky–Golay to determine the gas explosion characteristic parameter [2]. Since the sampling rate is 100 kHz in this study, the window of 75 points was used [3].

The stability and accuracy of the experimental apparatus is the key to evaluate the explosion suppression effect of porous materials. Prior to the experiment study on the explosion suppression effect of porous materials, the blank experiments without any obstruction A-1 and with thin iron hoop A-2 were performed three times, respectively. The maximum overpressure P_{max} measured in each group is compared, and the standard deviation of P_{max} at positions 4, 5, 6 and 7 is obtained, as shown in Fig. 7.3.

Figure 7.3 shows that the error of the P_{max} measured in the blank experiment is within 5%, the results implied that the reproducibility of experimental data was good. As indicated in Fig. 7.3a, the P_{max} is gradually increasing, and the P_{max} at position 7 was the greatest in the case of no obstacle. However, Fig. 7.3b shows the P_{max} was significantly different when the obstacle of a thin iron hoop was added, it can be found that the P_{max} at positions 4, 5, and 7 decreased, and the P_{max} at position 6 increases significantly, compared with Fig. 7.3a. Simultaneously, the oscillation of the overpressure in the spherical vessel is obvious, e.g., the pressure–time profiles of Group A-1 at position 4 as shown in Fig. 7.4.

Figure 7.4 shows the entire explosion process of the CH_4/Air mixture in the closed pipe-connected spherical vessel, in which the explosion characteristic parameters including the P_{max} and the maximum rate of pressure rise $(dp/dt)_{max}$ can be obtained (Standard Test Method for Explosion Parameters of Flammable Gases GB/T 803-2008; American Society for Testing and Materials (ASTM) ASTM E 681-04).

The overpressure-time curve shows a trend of rising first and then falling, the pressure begins to decrease due to the heat loss to the wall [4]. Oancea et al. [5] reported that the P_{max} could be given by Eq. (7.1).

$$P_{max} = k_{ad,V} \cdot P_0 - q_{tr} \frac{\gamma_e - 1}{V_0} \tag{7.1}$$

Fig. 7.4 Overpressure-time profiles at position 4 for Group A-1

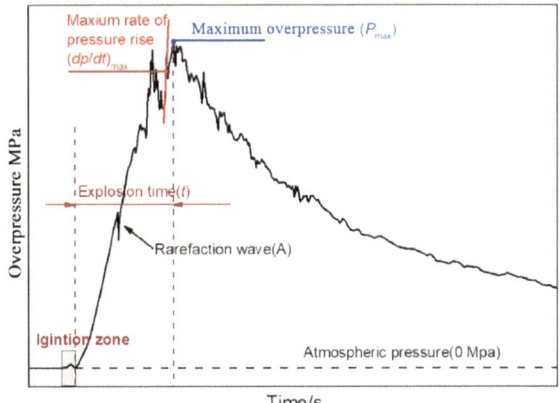

in which $k_{\mathrm{ad,V}} = P_{\mathrm{e}}/P_0$ is the adiabatic coefficient of the pressure development during an explosion, P_{e} is the adiabatic overpressure, P_0 is the actual overpressure, q_{tr} is the heat losses, γ_{e} is the specific heat ratio and V_0 is the volume of the experimental apparatus. Equation (7.1) indicates q_{tr} is the key factor for the evaluation P_{max}. Besides, there is a clear overpressure peak during the rapid development stage of each combustion explosion experiment (i.e., A in Fig. 7.3). Combustion and explosion occurred first in the spherical vessel, causing the overpressure to rapidly increase and propagate to the pipes before the explosion flames, finally, a rarefaction wave appeared during the explosion process [4, 6]. Furthermore, the rarefaction wave caused the overpressure of the spherical vessel to drop rapidly. When the flame front reached the connection part of the spherical vessel and the duct, the combustion of the gases was exacerbated because the flame front is unstable, and the overpressure increases again [7].

The parameters P_{max} and $(dp/dt)_{\mathrm{max}}$ of the blank experiment at positions 4, 5, 6, 7 are summarized in Table 7.3. The P_{max} of the blank experiment A-2 at position 4, 5 and 7 was reduced compared with that of no obstacle A-1, but the P_{max} at position 6 was increased to 0.5765 MPa and the $(dp/dt)_{\mathrm{max}}$ at position 4 reached to 19.5765 MPa s^{-1}.

Table 7.3 Explosion characteristic parameters of blank experiment

Item position	Test A-1		Test A-2	
	P_{max} (MPa)	$(dp/dt)_{\mathrm{max}}$ (MPa^2 s^{-1})	P_{max} (MPa)	$(dp/dt)_{\mathrm{max}}$ (MPa^2 s^{-1})
P4	0.4392 ± 0.0128	15.2453 ± 0.8927	0.4317 ± 0.0130	19.5765 ± 1.2987
P5	0.4445 ± 0.0025	4.8576 ± 0.6758	0.4099 ± 0.0138	5.1347 ± 0.3564
P6	0.4507 ± 0.0100	8.5654 ± 0.8076	0.5765 ± 0.0018	5.8871 ± 0.7543
P7	0.4643 ± 0.0129	11.3087 ± 0.9754	0.4393 ± 0.0130	10.7864 ± 0.9688

As indicated in Table 7.3, when the thin iron hoop was added, the relative over-pressure difference ($\Delta P = P_{\text{max p6}} - P_{\text{max p5}}$) between position 5 and position 6 is much greater than that without an obstacle. The reason for this change is that the explosion flames of A-1 propagating from the spherical vessel into the pipes can be normal along the pipe, resulting in the trend of the overpressure keeps rising slowly along with the positions. When the thin iron hoop was added to the experimental apparatus, it becomes an obstacle due to the inner diameter of the iron hoop is smaller than that of the pipes. When the flame interacts with the iron hoop, the flame front appears to be more turbulent and corrugated indicating that the surface area of the flame increases rapidly. A violent physical–chemical coupled activation occurs at the downstream of the iron hoop, thereby the combustion reaction rate is boosted. This may result in an overpressure enhancement [8–11]. Consequently, the P_{max} at position 6 is increased, and the $(dp/dt)_{\text{max}}$ at position 4 and position 5 is greater than A-1. These indicated that the thin iron hoop plays an important role in influencing the pressure propagation of the CH_4/Air mixture explosion.

7.1.3   Effect of Porous Materials on Explosion Suppression

As shown in Table 7.2 (Group B), in order to study on the characteristics of explosion suppression, six subcategories of the porous materials were selected to carry out. Figure 7.5a, b illustrated the overpressure-time curves at position 6 and the mean of maximum overpressure at position 4, 5, 6 and 7 for six subcategories of porous materials, respectively.

It is noticeable from Fig. 7.5a that the P_{max} of B-4 SiC 20 mm/10 PPI is the largest, and the P_{max} of B-6 Al_2O_3 10 mm/30 PPI is the smallest. Obviously, the one-layer porous materials exhibit a positive effect on the explosion suppression of CH_4/Air mixture compared to the blank experiment, where B-4 is negative. The porous material plays both positive and negative roles in affecting the explosion parameters. On one hand, it can achieve flame quenching and heat dissipation, showing a negative

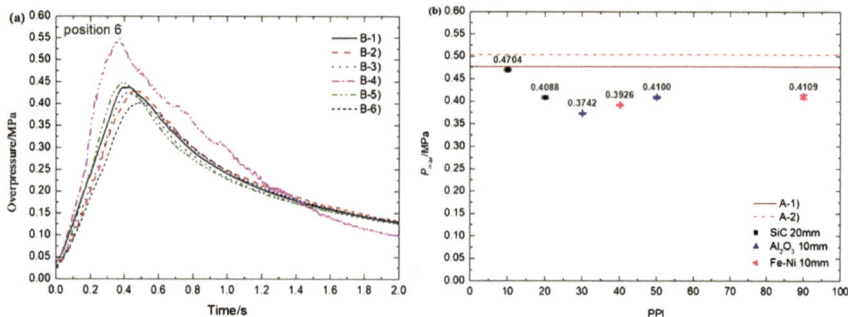

Fig. 7.5 Overpressure of one layer porous materials

effect on the explosion. On the other hand, it acts as a turbulence generator, under which the explosion process can be facilitated. As explosion waves arrive at the pore surface of porous materials from different angles, the reflected and scattered waves formed will exert vibrations of the air and porous struts, most of the explosion wave energy is consumed and converted into heat energy [12]. Moreover, the interconnectivity of the porous materials could provide a matrix of tortuous narrow passageways that the flam passes through, the flame is divided into large numbers of flame streams immediately because of the pores of porous material. The mean P_{max} of B-4 is significantly larger than the other ones, it's almost the same as the blank experiment with thin iron hoop. Besides, it can be found that the worse the explosion suppressive effect, the earlier the P_{max} reaches for the one-layer porous materials.

To further quantitatively analyze the effect of porous materials on explosion suppression, the characteristic parameter K (MPa^2 s^{-1}) is defined as the product of the P_{max} and the $(dp/dt)_{max}$, and the error value of K is obtained by the propagation of error [13]. Generally, the greater the K is the worse the explosion suppressive effect of the porous materials [14, 15]. The explosion intensity parameter K is summarized in Table 7.4, and the explosion intensity trend at positions 4, 5, 6 and 7 is shown in Fig. 7.6.

As illustrated in Table 7.4, it could be observed the K at position 4 are greater than at other ones. This is related to the space size of the combustion explosion, the larger space, the more adequate the combustion explosion reaction is [16]. As shown in Fig. 7.6, when the one-layer porous materials are added, the explosion intensity trend of the one-layer porous materials in the path of pressure propagation first decreased sharply, then increased slightly, and finally decreased slowly. Comparing the parameter K according to Table 7.4 and Fig. 7.6, it can be found that the explosion intensity trend of B-4 at position 4 is greater than that of A-1 and A-2, indicating that the explosion intensity suppression of B-4 is poor. The explosion intensity of B-1 Fe–Ni 10 mm/90 PPI and B-3 SiC 20 mm/20 PPI at position 4 is between A-1 and A-2, but these at positions 5, 6 and 7 is less than A-1 and A-2. Meanwhile, the explosion intensity of other one-layer porous materials is better than that of A-1 and A-2. The above results have shown that the one-layer porous materials can play a better explosion suppressive effect except B-4. Comprehensively, the optimal

Table 7.4 Explosion intensity parameters of one-layer porous materials

K (MPa^2 s^{-1}) Item	Position 4	Position 5	Position 6	Position 7
B test 1	8.2385 ± 1.6634	0.8142 ± 0.1771	1.0665 ± 0.1710	0.5638 ± 0.1645
B test 2	3.2856 ± 0.4651	0.6333 ± 0.0966	0.8280 ± 0.1843	0.7879 ± 0.1735
B test 3	8.7352 ± 0.4005	0.6555 ± 0.0523	0.6893 ± 0.0690	0.7244 ± 0.1554
B test 4	11.3740 ± 0.9406	1.8500 ± 0.3042	3.6184 ± 0.3396	4.8726 ± 0.6820
B test 5	4.1968 ± 0.2656	0.8696 ± 0.0731	1.0617 ± 0.0518	1.1153 ± 0.0063
B test 6	5.2776 ± 0.7046	0.5741 ± 0.0662	0.6541 ± 0.0797	0.7179 ± 0.1118

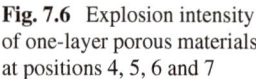

Fig. 7.6 Explosion intensity of one-layer porous materials at positions 4, 5, 6 and 7

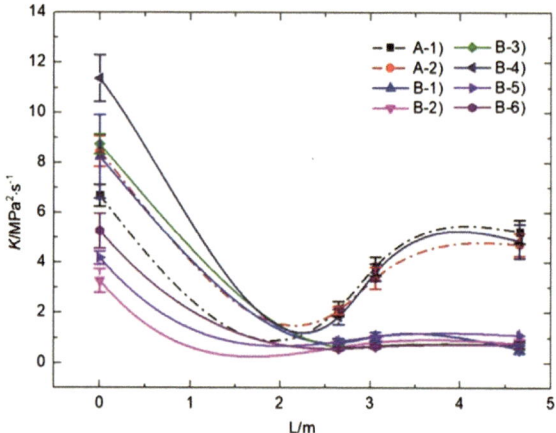

explosion suppressive effect of the one-layer porous materials is exhibited by B-2 Fe–Ni 10 mm/40 PPI.

The main reason for the poor explosion suppressive effect of B-4 is that the pore size is too large to give effective explosion suppression in the experiment. When the combustion flame passes through the porous materials, the larger pore size of the porous materials can not achieve quenching, but rather enhances the turbulence of flame combustion [17]. The pore size of the porous material B-1 is less than that of B-2 Fe–Ni 10 mm/40 PPI, but the effect of explosion intensity suppression is not as good as B-2, because the microstructure of the porous material can not only quench flames but also impede the propagation of overpressure [18, 19]. According to the law of conservation of momentum, the explosion flame through the porous media will cause viscosity loss (Darcy) and internal loss, manifested as flame quenching and pressure hindrance, which are closely related to the properties, thickness, pore size and combination mode of porous materials [20]. Therefore, the comprehensive explosion suppressive effect of B-2 on flame suppression and pressure propagation is superior to that of other one-layer porous materials.

The anti-sintering ability and impact resistance of the porous materials were analyzed by using a BX53M system microscope to observe the structure of one-layer porous materials after explosion suppressive experimental, and the results are shown in Fig. 7.7. It can be found that the surface morphology of the foam Fe–Ni porous material has obvious burning traces. The surface morphology of the SiC porous materials also has slight burning traces, while the Al_2O_3 porous material has a strong anti-sintering ability and no cauterization on the surface. And the black carbon produced by incomplete combustion remained in the internal structure of the porous material. The six subcategories of porous materials showed no internal structural damage, indicating that the impact resistance of porous materials is strong. Moreover, it is noticeable from Fig. 7.7 that the porous materials consist of pore and strut, and the spatial skeleton structure of porous material is three-dimensional network structure, with sound connectivity and large porosity. The large specific surface area

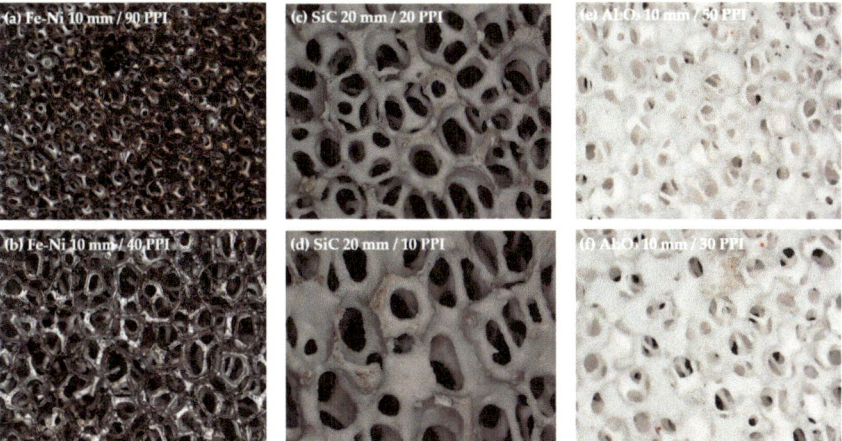

Fig. 7.7 BX53M system microscope images of porous materials showing anti-sintering the ability and impact resistance at 10 times magnification

hints increased heat dissipation. These characteristics play an important role in gas explosion suppression.

The porous materials were combined each other due to the difference between explosion suppression and anti-sintering ability. Consequently, the explosion suppressive characteristics of the two-layer composite porous material and the three-layer composite porous material in the pipe-connected spherical vessel were studied according to the combination mode in Table 7.2.

7.2 Interaction Mechanism of Wire Mesh and Venting on Gas Explosion

7.2.1 Experimental Apparatus and Methods

7.2.1.1 Experimental Apparatus

In this experiment, a combination of a spherical vessel and a duct were used, as shown in Fig. 7.8. The volume of the vessel was 113 L. The length of the pipe was 2000 mm and its inner diameter was 60 mm. The venting positions respectively located at the end of the duct and the top of the vessel, and the top venting diameter was 50 mm. A duct flange with a length of 250 mm was placed horizontally on the right side of the vessel, which was used to install the wire mesh and realize the connection with the venting duct. A polypropylene venting membrane was installed at the end of the duct to seal the duct. The premixed methane gas was prepared using the RCS2000-B automatic gas distribution system. A high-energy ignition system (KTD-A) was used to ignite the premixed gas (ignition energy (Q) = 2–6 J). The ignition position was the center of the vessel, which was 300 mm away from the vessel wall. Explosion flame propagation process was recorded by a high-speed camera with a maximum resolution of 1280 × 1024 pixel and the frame rate was 1000 fps. The pressure data was recorded by a high frequency pressure sensor. Further, the DEWE soft multichannel data acquisition system realized the program control and data acquisition.

The wire mesh material used in the experiment was 304 stainless steel. One-layer, three-layer, five-layer and 20-mesh, 40-mesh, 60-mesh wire mesh were used in the experiment, respectively. The metal wire mesh parameters used in the experiment are listed in Table 7.5. The wire mesh could withstand the high temperature and pressure of the methane explosion venting. In order to demonstrate adequate levels of repeatability, each experiment was repeated three times at least in this research to guarantee the experimental accuracy. According to the measured results, the maximum deviation of the repeatability experiment was 6.9%.

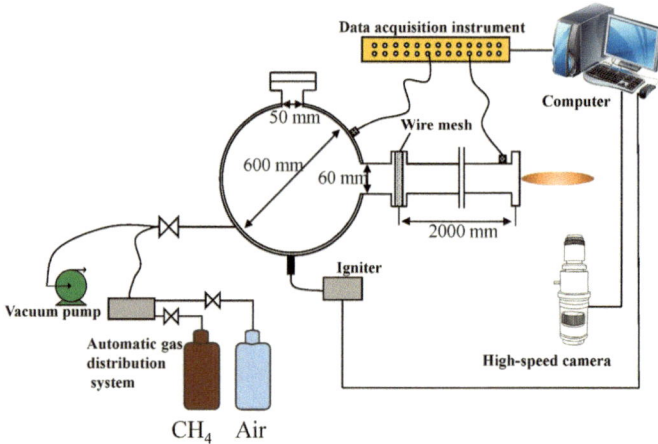

Fig. 7.8 Schematic diagram of the experimental equipment

Table 7.5 Properties of wire mesh types used in the tests

Number of wire mesh (N)	Number of holes per centimeter (N/cm)	Aperture size (mm)	Wire diameter (mm)	Volume fraction of metal (%)
20	7.87	0.949	0.314	0.389
40	15.7	0.439	0.194	0.478
60	23.6	0.302	0.154	0.574

7.2.1.2 Experimental Methods

The experiments for different kinds of explosion suppression parameters, explosion venting parameters and initial conditions were conducted, in order to analyse the interaction mechanism of the premixed gas explosion with wire mesh suppression and ducted venting. The experimental scheme is shown in Table 7.6.

The wire mesh parameters mainly included the number of layers and meshes. The explosion venting parameters mainly included the venting diameter and the rupture pressure (the pressure when the venting membrane ruptured). The initial conditions mainly included the initial concentration, initial pressure, ignition energy and venting position.

Table 7.6 Experimental scheme

Number	Explosion venting parameters		Initial conditions			
	Venting diameter (mm)	Rupture pressure (MPa)	Initial concentration (%)	Initial pressure (MPa)	Ignition energy (J)	Venting position
A1	30	0	9.50	0	6	Duct end
A2	40					
A3	50					
A4	60					
B1	60	0	9.50	0	6	Duct end
B2		0.05				
B3		0.10				
B4		0.15				
B5		0.30				
C1	60	0	6	0	6	Duct end
C2			8			
C3			9.50			
C4			12			
C5			14			
D1	60	0	9.50	0	6	Duct end
D2				0.01		
D3				0.02		
D4				0.03		
E1	60	0	9.50	0	2	Duct end
E2					3	
E3					4	
E4					5	
E5					6	

7.2.2   Effect of Mesh Parameters

7.2.2.1   Flame Structure

The effect of wire mesh number and layer on the flame structure outside the explosion vent is shown in Fig. 7.9. The three symbols represent three different layers of metal wire mesh, the number of wire mesh layers is one, three, and five respectively. It is observed that the external flame length of the explosion venting vessel decreases successively with an increase in the wire mesh number and layer. The aperture and opening rate of the metal wire mesh decreased as the wire mesh number and layer increased, which resulted in the increase of discretization degree. The increase of the contact area between the flame and the metal resulted the increase in the heat loss of the flame. The high heat loss of the flame resulted in the smaller propagation distance of the flame passing through the wire mesh, which means that the suppression effect of flame propagation was better. During the experiment, the vessel and duct were filled with premixed gas. In the case of 5 layers of 60 mesh steel wire mesh, there was no flame outside the venting port, which meant that the premixed gas behind the wire mesh was not ignited. This also indicated that the flame was inhibited by the wire mesh.

Fig. 7.9 Explosion venting flame structure under different wire mesh suppression conditions

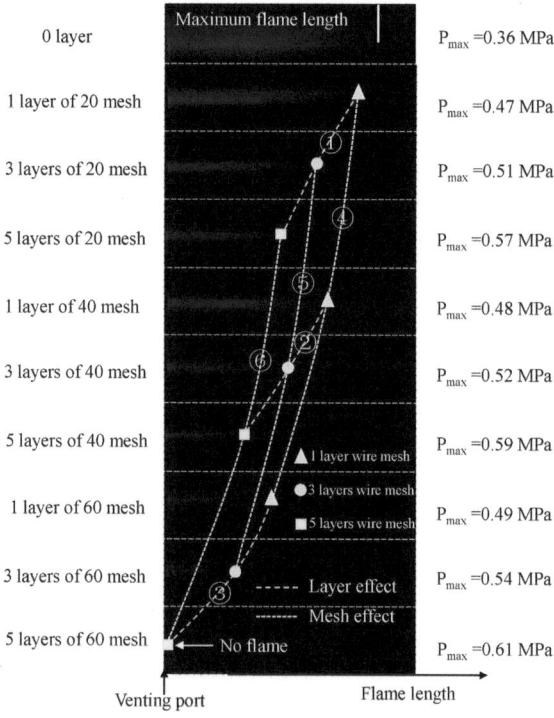

7.2.2.2 Internal Pressure

Figures 7.10 and 7.11 present the variation curves of P_{max} inside the vessel with the increasing of wire mesh number and layer. The figures indicate that the P_{max} inside the vessel was increased sequentially with an increase in the wire mesh number and layer [21, 22]. With the number of layers increased from one to five, the P_{max} of three kinds of wire meshes under explosion suppression conditions were increased by 0.10 MPa, 0.11 MPa and 0.12 MPa, respectively. Similarly, with an increase in the number of mesh, the P_{max} of the three kinds of the layers under explosion suppression conditions were increased by 0.02 MPa, 0.03 MPa and 0.04 MPa, respectively. It was observed that the effect of the layers was more significant than that of mesh number, and the P_{max} appeared the maximum under the working condition of five layers and 60 mesh number. This is because the metal wire mesh had a dense pore structure. The increase in the number of layers and meshes led to a decrease in the pore size and mesh rate, which resulted in the heat inside the vessel could not be released quickly owing to the decrease in mesh rate and the metal volume fraction was increased. Additionally, the venting of the flame was affected by the reverse pressure wave due to the obstruction of the metal wire mesh during the venting process [23–25]. In the process of flame propagation, the pressure wave would be prior to the flame propagation. During the interaction between the precursor shock wave and the wire mesh, one part of the shock wave would pass through the wire mesh. And the other part would be blocked by the vessel wall and the wire mesh, which resulted in backward propagation of shock wave and touching with the flame surface. The latter was called reverse shock wave [26, 27]. With an increase in the number of wire mesh and layer, the reverse pressure wave was also increased. And the internal pressure venting rate of the vessel slowed down, resulting in a sharp increase in the internal pressure of the vessel. Therefore, the P_{max} inside the vessel showed an increasing variation trend with an increase in the wire mesh number and layer.

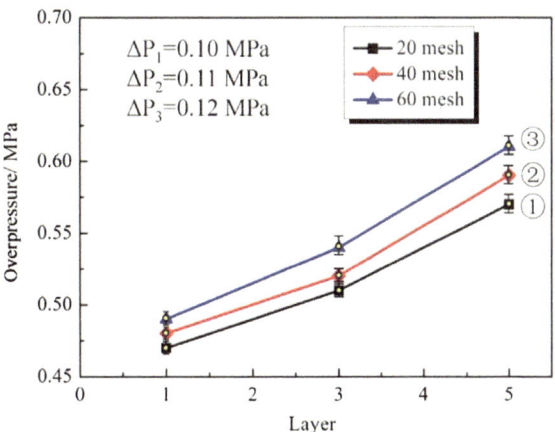

Fig. 7.10 Change of the P_{max} inside the vessel under the different wire mesh layers

Fig. 7.11 Change of the P_{max} inside the vessel under the different wire meshes

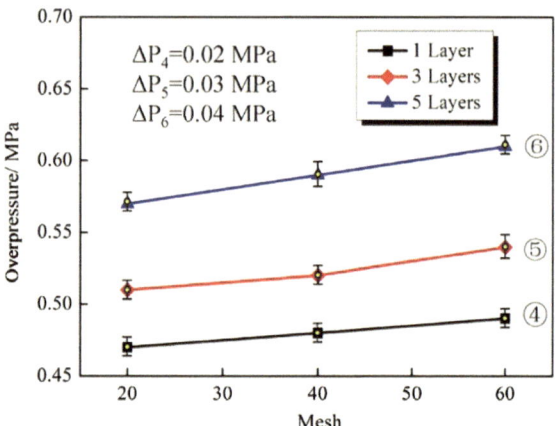

7.2.3 Effect of Venting Parameters

7.2.3.1 Venting Diameter

In order to achieve an optimal combination of explosion suppression and venting, it is not only necessary to ensure that there is no venting flame outside the vessel but also a small overpressure exists inside the vessel during practical applications. Therefore, to investigate the effect of the venting diameter on the internal pressure of the vessel, a 5-layer 60-mesh wire (without venting flame) was adopted for explosion suppression working condition. Figure 7.12 presents the effect of the venting diameter on the overpressure curves inside the vessel. The P_{max} showed an increasing trend with the decrease of venting diameter (from 0.610 to 0.690 MPa). With the increase of venting diameter (from 30 mm to 60 mm), the standard deviation of the P_{max} of repeated experiments were 0.029, 0.026, 0.033 and 0.034, respectively. It can be seen that the measured pressure can ensure the accuracy of the experiment. Under the 30 mm venting diameter, the overpressure inside the vessel reached the maximum value ($P_{max} = 0.690$ MPa). It is because that the venting rate of explosion flame was increased obviously with an increase in the venting diameter, which resulted in the reduce of the heat accumulation rate inside the vessel [28, 29]. Additionally, the reflection effect of the overpressure wave was enhanced with the decrease of venting diameter when it passed through the vent port. The turbulence and heat accumulation were also increased [30]. As a result, the P_{max} inside the vessel was increased with a decrease in the venting diameter, and its appearance moment was decreased gradually.

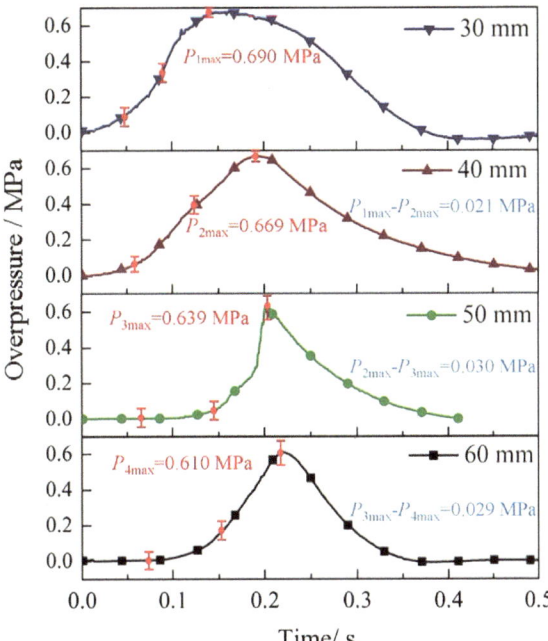

Fig. 7.12 Variation curves of the overpressure inside the vessel under different venting diameters

7.2.3.2 Rupture Pressure

The overpressure inside the vessel had a minimum value as the venting diameter was 60 mm through the above analysis. Under this venting diameter, 5-layer 60-mesh wire meshes were selected to research the effect of the rupture pressure on the overpressure. Figure 7.13 presents the effect of rupture pressure on the overpressure curves inside the vessel. As shown in Fig. 7.13, the P_{max} inside the vessel was increased with an increase in the rupture pressure (from 0.610 to 0.655 MPa) and its appearance moment was reduced [31, 32]. Similarly, $(dp/dt)_{max}$ also exhibited an increasing trend. Compared with no venting membrane, the P_{max} inside the vessel was increased by 1.8% and 7.4%, and $(dp/dt)_{max}$ was increased by 141% and 512% as the rupture pressure was 0.10 MPa and 0.30 MPa, respectively. With the increase of rupture pressure (from 0 MPa to 0.30 MPa), the standard deviations of the P_{max} were 0.034, 0.034, 0.035, 0.036 and 0.036, respectively. The overpressure inside the vessel could not be released instantaneously due to the increase in rupture pressure, resulting in the rapid accumulation of heat and then showing an increase trend of pressure inside the vessel.

Fig. 7.13 Overpressure
variation inside the vessel
with time under different
rupture pressures

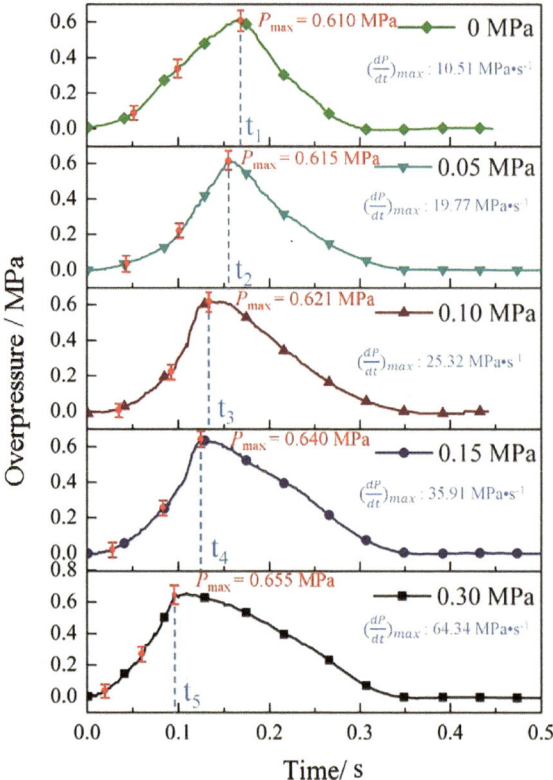

7.2.4 Effect of Initial Conditions

7.2.4.1 Initial Concentration

A plot of the variation curves of the overpressure inside the vessel with time for
different initial concentrations (concentration of methane) is shown in Fig. 7.14.
Methane was selected as fuel and the initial concentration referred to the volume
concentration of methane during experiments. Five kinds of methane concentrations
(6%, 8%, 9.5%, 12% and 14%) were selected during the explosion venting experiments, and the corresponding equivalence ratios (Φ) were 0.61, 0.83, 1, 1.06, 1.30
and 1.55, respectively. The overpressure inside the vessel first increased and then
decreased with an increase in methane concentration within the explosive limit [33–
35]. During the explosion venting process, the overpressure reached the maximum
($P_{max} = 0.610$ MPa) as the methane concentration was 9.5% ($\Phi = 1$). It was because
that methane could react completely with air. As $\Phi < 1$, the explosion intensity
was decreased owing to the insufficient fuel (from 0.610 to 0.114 MPa). As $\Phi > 1$,
the lack of combustion-supporting gas (oxygen) required for the reaction was one
of the reasons for the decrease of explosion intensity (from 0.610 to 0.168 MPa).

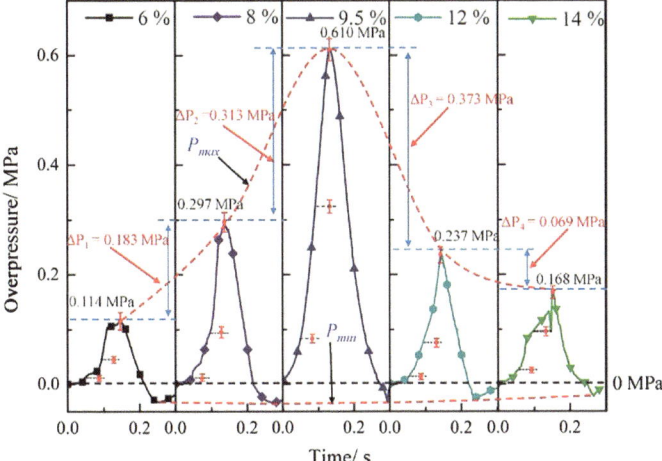

Fig. 7.14 Variation curves of overpressure inside the vessel with time under different initial concentrations of methane

Simultaneously, the pressure decreased sharply after reaching the P_{max}. With the increase of initial concentration (from 6% to 14%), the standard deviations of the P_{max} were 0.005, 0.011, 0.030, 0.007 and 0.007, respectively. Meanwhile, the pressure decreased to a negative value in Fig. 7.14. With the increasing initial pressure, the negative pressure inside the vessel was increased. After the pressure inside the vessel was reduced to atmospheric pressure, the explosive gas continued to be vented to the outside the vessel due to inertia, which leaded to the negative pressure inside the vessel. The explosion strength was enhanced and the greater the inertia of explosion venting with the increasing initial pressure, which resulted in the greater the negative pressure inside the vessel. This phenomenon was also described in previous research works [36–40].

7.2.4.2 Initial Pressure

The overpressure inside the vessel can be affected obviously by the initial pressure inside the vessel during the explosion venting process. A plot of the variation curves of the overpressure inside the vessel with time for different initial pressures is shown in Fig. 7.15.

As shown in Fig. 7.15, the P_{max} inside the vessel was proportional to the initial pressure. The P_{max} was increased significantly with the increase of initial pressure (from 0.610 to 1.132 MPa). Compared with atmospheric pressure, the P_{max} at initial pressures of 0.01 MPa, 0.02 MPa and 0.03 MPa were increased by 13.61%, 38.36% and 85.57%, respectively. With the increase of initial pressure from (0 MPa to 0.03 MPa), the standard deviations of the P_{max} were 0.015, 0.021, 0.019 and 0.024, respectively. According to the ideal gas law, the higher the initial pressure results in a

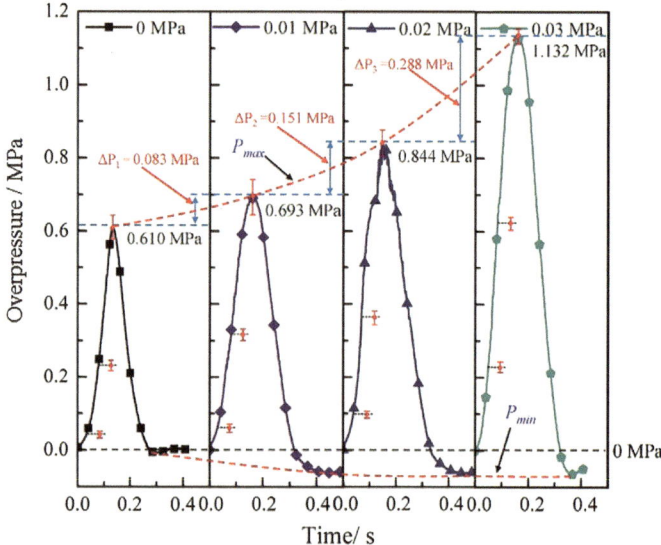

Fig. 7.15 Overpressure variation inside the vessel with time under different initial pressures

shorter distance between the premixed gas molecules, which resulted in the increase of molecules number per unit volume. Hence, the enhancement in molecular activity resulted in an increase in the collisions possibility, which accelerated the methane explosion reaction rate and increased the explosion intensity [41–43]. After ignition, the shock wave was vented to the outside the vessel through the vent port. However, the exhaust volume exceeded the amount of gas produced by combustion in the later stage of explosion venting, and the pressure inside the vessel gradually decreased to atmospheric pressure. And the gas generated by the explosion continued to vent due to inertia, resulting in a small negative pressure. Meanwhile, the explosion intensity was increased gradually with the increasing initial pressure, and the inertia of explosion venting got more intense, resulting the obvious negative pressure inside the vessel.

7.2.4.3 Ignition Energy

A plot of the variation curves of the pressure inside the vessel with time for different ignition energies is shown in Fig. 7.16. It can be observed that the P_{max} was increased with an increase in ignition energy under combined the wire mesh explosion suppression and the ducted venting (from 0.298 to 0.610 MPa). As 2 J < Q < 4 J, the effect of ignition energy on the P_{max} was significant (from 0.298 to 0.517 MPa). And then the effect was decreased as the ignition energy between 4 and 6 J (from 0.517 to 0.610 MPa) [44, 45]. Compared with 2 J, P_{max} were increased by 40.60%, 73.48%, 89.93% and 104.70% respectively with increasing ignition energy. With the increase

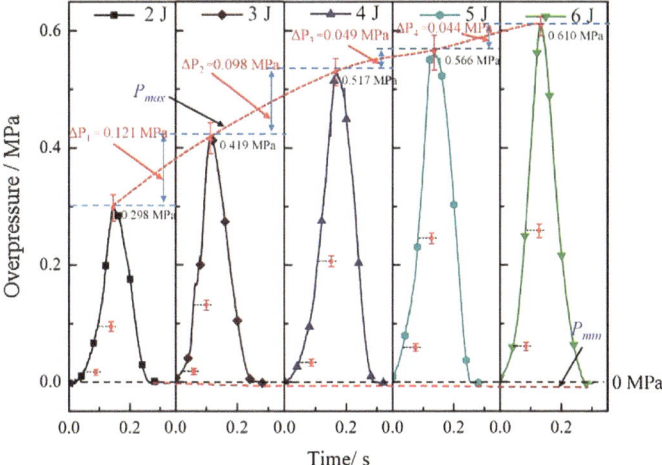

Fig. 7.16 Overpressure variation inside the vessel with time under different ignition energies

of ignition energy (from 2 J to 6 J), the standard deviations of the P_{max} were 0.012, 0.019, 0.023, 0.023 and 0.017, respectively. It is because that the explosion process was a complex multistep chain reaction process. The ignition process of methane explosion was a chain reaction process composed of several basic reactions. The occurrence of methane explosion reaction needed a certain amount of energy to break the chemical bond and produce free radicals. The chemical reaction could be initiated as the molecules whose ignition energy exceeded a certain value (activation energy). More free radicals were produced with increasing ignition energy, and the number of activated molecules in unit volume was increased. This also leaded to an increase in the energy that initiated the chain reaction. The more radical participated in the chain reaction, which resulted in the more complete the explosion and the stronger the explosion intensity. Macroscopically, P_{max} was increased with increasing ignition energy [46].

7.2.4.4 Venting Position

As investigating the effect of venting position, the methane concentration, initial pressure and ignition energy were 9.5%, 0 MPa and 6 J. The 5-layer 60-mesh wire mesh structure was adopted. Figure 7.17 presents the variation curves of the overpressure inside the vessel and the duct end under different venting positions. The venting position has a significant effect on overpressure [47]. After ignition, the overpressure underwent a rising process, and then the overpressure was vented outside the vessel from the vent port. However, the explosion energy produced by the combustion reaction inside the vessel was greater than the vented energy, which leaded to the continuous increase of the pressure inside the vessel. Then, the explosion intensity

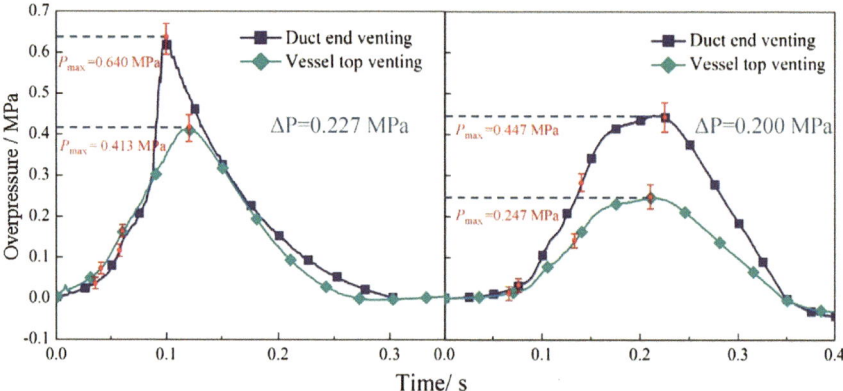

Fig. 7.17 Variation curves of overpressure inside the vessel and duct end with time under different venting positions

was weakened with the decrease of fuel inside the vessel, the overpressure was not enough to offset the pressure vented from the vent port, resulting in the decrease of overpressure [48]. Compared with the duct end venting, the P_{max} inside the vessel and the duct end were reduced by 35.5% and 44.7% respectively under the vessel top venting. The standard deviations of the P_{max} inside the vessel and at the end of the pipe under the vessel top venting were 0.014 and 0.009, respectively. And the standard deviations under the duct end venting were 0.031 and 0.015, respectively. It is because that the flame was accelerated inside the duct due to the unrestricted flame propagation space, resulting in a significant increase in pressure. Compared with the horizontal duct end venting, in the case of top venting, the overpressure wave was easier to vent to the outside the vessel due to buoyancy. And the time for the pressure shock wave to reach the vent port was also greatly shortened. Hence, the venting rate of pressure shock wave was increased and the rate of heat accumulation inside the vessel was reduced significantly.

References

1. Sun, Z. Y. (2018). Explosion pressure measurement of 50% H_2–50% CO synthesis gas–air mixtures in various turbulent ambience. *Combustion Science and Technology, 190*(6), 1007–1022.
2. Savitzky, A. (1964). Smoothing and differentiation of data by simplified least squares procedures. *Analytical Chemistry, 36*(8), 1627–1639.
3. Razus, D., Oancea, D., Chirila, F., et al. (2003). Transmission of an explosion between linked vessels. *Fire Safety Journal, 38*(2), 147–163.
4. Razus, D. M., & Krause, U. (2001). Comparison of empirical and semi-empirical calculation methods for venting of gas explosions. *Fire Safety Journal, 36*(1), 1–23.

5. Oancea, D., Gosa, V., Ionescu, N. I., et al. (1985). An experimental method for the measurement of the adiabatic maximum pressure during an explosive gaseous combustion. *Revue Roumaine de Chimie, 30*(9–10), 767–776.

6. Ye, J. F., Jiang, X. H., Jia, Z. W., et al. (2004). Experimental investigations of external second-explosion induced by vented explosion. *Explosion and Shock Waves, 24*(4), 356–362.

7. Wen, X., Xie, M., Yu, M., et al. (2013). Porous media quenching behaviors of gas deflagration in the presence of obstacles. *Experimental Thermal and Fluid Science, 50*, 37–44.

8. Zhang, J., Sun, Z., Zheng, Y., et al. (2012). Coupling effects of foam ceramics on the flame and shock wave of gas explosion. *Safety Science, 50*(4), 797–800.

9. Oh, K. H., Kim, H., Kim, J. B., et al. (2001). A study on the obstacle-induced variation of the gas explosion characteristics. *Journal of Loss Prevention in the Process Industries, 14*(6), 597–602.

10. Lee, J. H. S. (1983). Gas cloud explosion—Current status. *Fire Safety Journal, 5*(3–4), 251–263.

11. Wang, L. Q., Ma, H. H., Shen, Z. W., et al. (2019). The influence of an orifice plate on the explosion characteristics of hydrogen-methane-air mixtures in a closed vessel. *Fuel, 256*, 115908.1-115908.6.

12. Nie, B., He, X., Zhang, R., et al. (2011). The roles of foam ceramics in suppression of gas explosion overpressure and quenching of flame propagation. *Journal of Hazardous Materials, 192*(2), 741–747.

13. Verma, A. K., Srividya, A., & Rana, A. (2015). Use of stochastic petrinets in modeling of safety device inspection interval problem. *International Journal of Systems Assurance Engineering and Management, 6*(2), 1–7.

14. Cui, Y. Y., Wang, Z. R., Zhou, K. B., et al. (2016). Effect of wire mesh on double-suppression of CH₄/air mixture explosions in a spherical vessel connected to pipelines. *Journal of Loss Prevention in the Process Industries, 45*, 69–77.

15. Zhang, S., Wang, Z., Zuo, Q., et al. (2016). Suppression effect of explosion in linked spherical vessels and pipelines impacted by wire-mesh structure. *Process Safety Progress, 35*(1), 68–75.

16. Zhang, K., Wang, Z. R., Yan, C., et al. (2017). Effect of size on methane-air mixture explosions and explosion suppression in spherical vessels connected with pipes. *Journal of Loss Prevention in the Process Industries, 49*, 785–790.

17. Korzhavin, A. A., Bunev, V. A., Babkin, V. S., et al. (2005). Selective diffusion during flame propagation and quenching in a porous medium. *Combustion, Explosion, and Shock Waves, 41*(4), 405–413.

18. Fothergill, C. E., Chynoweth, S., Roberts, P., et al. (2003). Evaluation of a CFD porous model for calculating ventilation in explosion hazard assessments. *Journal of Loss Prevention in the Process Industries, 16*(4), 341–347.

19. Koester, G. E. (1997). *Propagation of wave-like unstabilized combustion fronts in inert porous media* [Master's thesis]. The Ohio State University.

20. Birk, A. M. (2008). Review of expanded aluminum products for explosion suppression in containers holding flammable liquids and gases. *Journal of Loss Prevention in the Process Industries, 21*(5), 493–505.

21. Xianfeng, C., Zhao, Q., Huaming, D., et al. (2018). Effect of metal mesh on the flame propagation characteristics of wheat starch dust. *Journal of Loss Prevention in the Process Industries, 55*, 107–112.

22. Zhang, B. (2016). The influence of wall roughness on detonation limits in hydrogen–oxygen mixture. *Combustion and Flame, 169*, 333–339.

23. Zhang, B., & Liu, H. (2017). The effects of large scale perturbation-generating obstacles on the propagation of detonation filled with methane-oxygen mixture. *Combustion and Flame, 182*, 279–287.

24. Cao, X., Ren, J., Zhou, Y., et al. (2015). Suppression of methane/air explosion by ultrafine water mist containing sodium chloride additive. *Journal of Hazardous Materials, 285*, 311–318.

25. Dong, C., Bi, M., & Zhou, Y. (2012). Effects of obstacles and deposited coal dust on characteristics of premixed methane–air explosions in a long closed pipe. *Safety Science, 50*(9), 1786–1791.

26. Cao, Y., Li, B., & Gao, K. (2018). Pressure characteristics during vented explosion of ethylene-air mixtures in a square vessel. *Energy, 151*, 26–32.
27. Xing, H., Xu, Q., Song, X., et al. (2020). The effects of vent area and ignition position on pressure oscillations in a large L/D ratio duct. *Process Safety and Environmental Protection, 135*, 166–170.
28. Cao, Y., Guo, J., Hu, K., et al. (2017). Effect of ignition location on external explosion in hydrogen–air explosion venting. *International Journal of Hydrogen Energy, 42*(15), 10547–10554.
29. Zhang, S., Tang, Z., Li, J., et al. (2019). Effects of equivalence ratio, thickness of rupture membrane, and vent area on vented hydrogen–air deflagrations in an end-vented duct with an obstacle. *International Journal of Hydrogen Energy, 44*(47), 26100–26108.
30. Cao, Y., Gao, K., Li, B., et al. (2023). Influence of vent size and vent burst pressure on vented ethylene-air explosion: Experimental and numerical study. *Process Safety and Environmental Protection, 170*, 297–309.
31. Wang, S., Yan, Z., Li, X., et al. (2020). The venting explosion process of premixed fuel vapour and air in a half-open vessel: An analysis of the overpressure dynamic process and flame evolution behaviour. *Fuel, 268*, 117385.
32. Liang, Z. (2017). Scaling effects of vented deflagrations for near lean flammability limit hydrogen-air mixtures in large scale rectangular volumes. *International Journal of Hydrogen Energy, 42*(10), 7089–7103.
33. Rui, S., Guo, J., Li, G., et al. (2018). The effect of vent burst pressure on a vented hydrogen–air deflagration in a 1 m^3 vessel. *International Journal of Hydrogen Energy, 43*(45), 21169–21176.
34. Guo, J., Li, Q., Chen, D., et al. (2015). Effect of burst pressure on vented hydrogen-air explosion in a cylindrical vessel. *International Journal of Hydrogen Energy, 40*(19), 6478–6486.
35. Tascón, A., Ruiz, Á., & Aguado, P. J. (2011). Dust explosions in vented silos: Simulations and comparisons with current standards. *Powder Technology, 208*(3), 717–724.
36. Tascón, A., & Aguado, P. J. (2015). CFD simulations to study parameters affecting dust explosion venting in silos. *Powder Technology, 272*, 132–141.
37. Cen, K., Song, B., Huang, Y., et al. (2017). CFD simulations to study parameters affecting gas explosion venting in compressor compartments. *Mathematical Problems in Engineering, 2017*, 1090561.
38. Poli, M., Grötz, R., & Schröder, V. (2012). An experimental study on safety-relevant parameters of turbulent gas explosion venting at elevated initial pressure. *Procedia Engineering, 42*, 90–99.
39. Zhang, B., & Liu, H. (2019). Theoretical prediction model and experimental investigation of detonation limits in combustible gaseous mixtures. *Fuel, 258*, 116132.
40. Cheng, J., Zhang, B., Liu, H., et al. (2021). The precursor shock wave and flame propagation enhancement by CO_2 injection in a methane-oxygen mixture. *Fuel, 283*, 118917.
41. Li, H., Guo, J., Tang, Z., et al. (2019). Effects of ignition, obstacle, and side vent locations on vented hydrogen–air explosions in an obstructed duct. *International Journal of Hydrogen Energy, 44*(36), 20598–20605.
42. Pascaud, J. M. (2016). Influence of a dispersed ignition in the explosion of two-phase mixtures. *Combustion Science and Technology, 188*(10–12), 1719–1740.
43. Di Benedetto, A., Garcia-Agreda, A., Russo, P., et al. (2012). Combined effect of ignition energy and initial turbulence on the explosion behavior of lean gas/dust-air mixtures. *Industrial and Engineering Chemistry Research, 51*(22), 7663–7670.
44. Luo, X., Wang, C., Rui, S., et al. (2020). Effects of ignition location, obstacles, and vent location on the vented hydrogen-air deflagrations with low vent burst pressure in a 20-foot container. *Fuel, 280*(2), 118677.
45. Cao, W., Li, W., Yu, S., et al. (2021). Explosion venting hazards of temperature effects and pressure characteristics for premixed hydrogen-air mixtures in a spherical container. *Fuel, 290*(2), 120034.
46. Cao, W., Liu, Y., Chen, R., et al. (2021). Pressure release characteristics of premixed hydrogen-air mixtures in an explosion venting device with a duct. *International Journal of Hydrogen Energy, 46*(12), 8810–8819.

47. Yang, K., Hu, Q., Sun, S., et al. (2019). Research progress on multi-overpressure peak structures of vented gas explosions in confined spaces. *Journal of Loss Prevention in the Process Industries, 62*, 103969.
48. Dai, H., Wang, X., Chen, X., et al. (2020). Suppression characteristics of double-layer wire mesh on wheat dust flame. *Powder Technology, 360*, 231–240.

Chapter 8
Suppression Effects of Ultrafine Water Mist on Gas Explosion

8.1 Experimental Apparatus and Methods

The experimental apparatus was consisted of an explosion vessel, a gas supply system, a mist generation system, an ignition system, and a process control and data acquisition system, as shown in Fig. 8.1. The dimension of the explosion vessel is 910 mm × 150 mm × 150 mm and the design pressure is 1.50 MPa. Two tempered glasses were installed in the front and back sides of the explosion vessel for the visualization of the flame evolution process. A pair of ignition electrodes with a gap of 5 mm was set 8 cm apart from the bottom of the vessel to activate the explosion. The high-frequency data acquisition card (PCI8348AJ), with high-speed parallel analog inputs and programmable digital outputs, was adopted in the process control and data acquisition system to start the ignition and pressure acquisition in proper sequence. A 50 kHz piezoresistive pressure sensor (MDHF20) with a dynamic responding time of 1 ms was installed in the middle of the explosion vessel to acquire the pressure history. A high-speed camera (FASTCAM SA4) was used to record the flame propagation process. The frame rate and saving format (videos or photos) of the shooting process were controlled by program.

The ultrasonic atomization system mainly includes an ultrasonic fogger unit, an atomization cup (80 mm × 80 mm × 150 mm) with four guide fins, a transformer, etc., as is shown in Fig. 8.2. The system was located at the top of the explosion vessel. Once connected the power, the mist was generated inside the vessel directly. With the mist amount increased, the ultrafine water mist spread into the whole vessel through the outlets uniformly distributed in the sidewall of the atomization cup. The ultrafine mist fell very slowly after escaping from the atomizing cup and wouldn't settle a lot under the action of gravity and drag forces. The amount of atomization was greater significantly than the amount of settled mist. Meanwhile, in order to avoid the effect of the visual windows on the diameter mensuration process, the diameter of the mist was measured before the mist was injected into the vessel. The generating rate of the mist was 1.146 g/min which was measured by the precision balance, and the mist size

Z. Wang and X. Cao, *Gas Explosion and Its Protection Technology in Process Industries*, https://doi.org/10.1007/978-981-96-3121-6_8

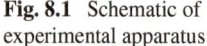
Fig. 8.1 Schematic of experimental apparatus

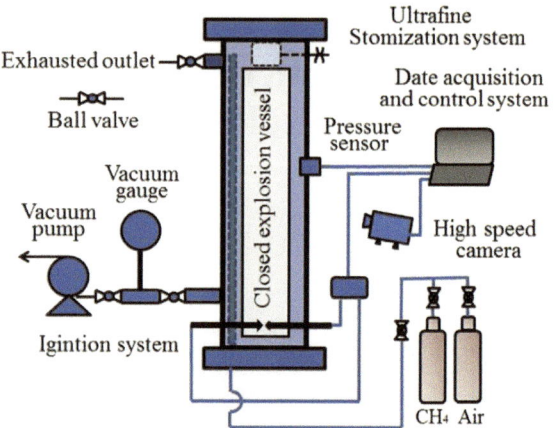

was 10.03 μm which was measured by a phase Doppler particle analyzer. Before the outlet, there laid a stainless-steel mesh guarding against the large diameter droplet splashing into the vessel and causing large disturbance to the flame front.

Based on Dalton's law of partial pressure, the premixed gas of 6.5, 8, 9.5, 11, and 13.5% methane concentration was prepared in the combustion chamber representatively. According to generation rate of the mist and the scheduled time, the suppression experiments were conducted under eight spraying concentrations (i.e., 56, 112, 168, 224, 280, 336, 448, 560, and 840 g/m^3). Initial pressure was 0.1 MPa before experiments. After ignition, the flame propagated upward and the pressure began to rise rapidly. The explosion flame and overpressure were recorded by high-speed camera and high frequency pressure sensor. To guarantee the accuracy of experimental data, each suppression experiment was repeated 4 times at least.

Fig. 8.2 Schematic of ultrasonic atomization system

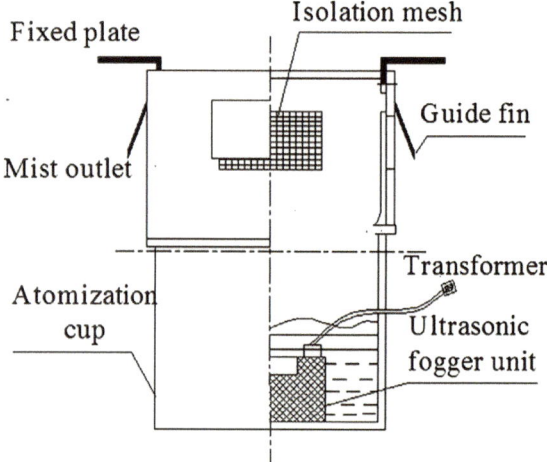

8.2 Visualization of Flame Propagation Process

8.2.1 Relationship of Pressure Rising and Flame Propagation

Figure 8.3 presents the relationship between pressure rising rate, flame propagation velocity and flame front structure of 9.5% methane explosion under no spraying. At 0 ms, the premixed gas was ignited and the flame developed in the form of sphere. The flame expanded outwards freely, without effects of the sidewalls. The pressure rising rate history increased slowly due to slower flame propagation velocity in the initial stage (t = 0–32 ms). With the increase of the diameter, temperature and pressure of combustion wave in the flame front got higher, which made the combustion products expand rapidly. The flame began to accelerate after 32 ms due to the increase of the flame surface area, and the pressure rising rate history increased rapidly. At 55 ms, the maximum flame propagation velocity (Vmax) appeared which indicated that the flame skirt touched the sidewalls [1], and the subsequent flame deceleration was observed due to the flame surface loss. During this stage, the flame propagated in the form of finger (t = 32–88 ms). The curvature radius of flame front increased gradually after 55 ms with the flame propagating velocity decreasing sharply. As the pressure rising rate reached the first peak value, flame propagation velocity was minimum value and the flame front changed into the "plane" flame (t = 88 ms). The accelerating propagation of combustion wave resulted in the first accelerating rise of pressure rising rate history and the appearance of first peak value. Subsequently, the flame near the sidewall began to accelerate and the flame started to invert, which indicated that the "tulip" flame started to appear. With the formation of the "tulip" flame, the flame velocity and pressure began to increase again due to the second growth in the flame surface area (t = 140–208 ms). Moreover, the velocity history appeared oscillation variation in the later period of flame propagation [2]. It is deduced that the combustion wave would be affected by reverse action of pressure wave (or acoustic wave) due to the blocking of the vessel end. At 208 ms, the pressure rising rate increased to the maximum value as the flame reached the brightest intensity. Then the flame began to extinguish and the pressure rising rate decreased rapidly.

8.2.2 Flame Front Structure

As is shown in Fig. 8.3, the explosion flame propagation process experienced "sphere" flame → "finger" flame → "plane" flame → "tulip" flame. After the "plane" flame formed, the flame moving velocity near the sidewall began to accelerate and the flame velocity in the central axis slow down suddenly. It means that the "tulip" flame formed gradually. Therefore, the sudden acceleration of flame propagation near the sidewall was the signal of "tulip" flame appearance. The flame front position at this moment was defined as the appearance height of the "tulip" flame. The

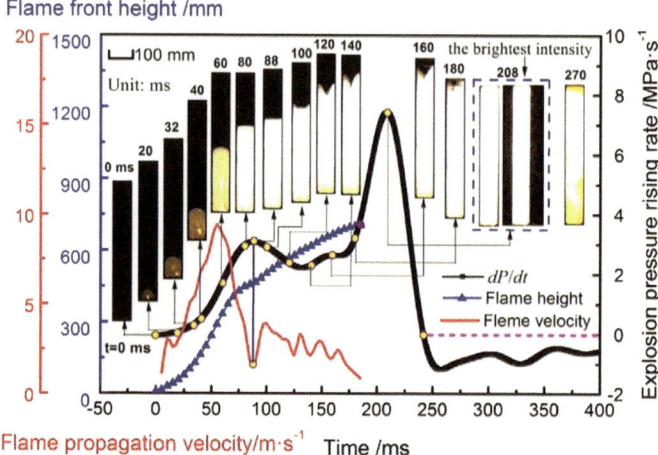

Fig. 8.3 The relationship between pressure rising and flame propagation

solid red lines stand for the positions of the flame front as the flame near the sidewall began to sudden accelerate after "plane" flame in Fig. 8.4. Figure 8.4a–c presents the appearance height and corresponding moment of the "tulip" flames of 8, 9.5, and 11% methane explosions with increasing spraying concentration. The appearance height of the "tulip" flame was increased and corresponding moment was delayed obviously.

Figure 8.5 shows the effect of spraying concentration on the percentage of "tulip" flame appearance height to the total height of the vessel, and on the percentage of the "tulip" flame appearance time after ignition to the total time of flame propagation from the bottom to the top of the vessel. It can be seen that the height percentage and the time percentage were increased obviously with increasing spraying concentration. For 6% and 13.5% methane explosions, the "tulip" flame would not appear when spraying concentrations exceeded 112 g/m^3 and 168 g/m^3, respectively. As methane concentration was far away from the stoichiometric ratio, the combustion wave was weakened easily by water mist due to the weaker explosion intensity. It is concluded that the appearance of "tulip" flame was affected by the explosion intensity and spraying concentration.

The formation of "tulip" flame could be attributed to the interactions among the flame front, the flame-induced reverse flow and the vortices behind the flame front. The existences of the reverse flow and the vortices in burned gas have been demonstrated by the PIV images in Bogdan's experiments [3]. And the evolution process of explosion flow field was also described in the numerical simulation [4]. During the flame propagation, the reverse flow appeared behind the flame front and its appearance was prior to that of two symmetrical vortices near the bottom end. The reverse flow became stronger with the flame accelerating propagation, which resulted that the two symmetrical small vortices were initiated and moved toward the flame front with a higher velocity relative to the flame front, then developed to two

Fig. 8.4 Effect of spraying concentration on the "tulip" flame appearance. **a** % CH₄, **b** 9.5% CH₄, **c** 11% CH₄

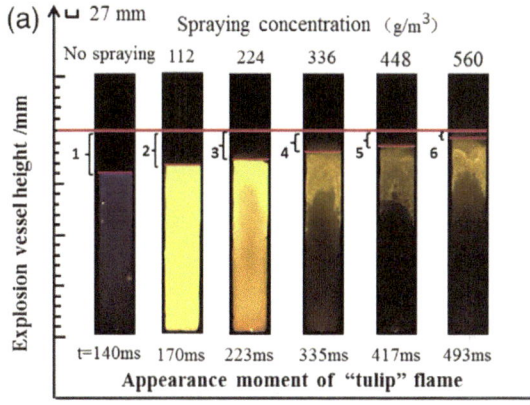

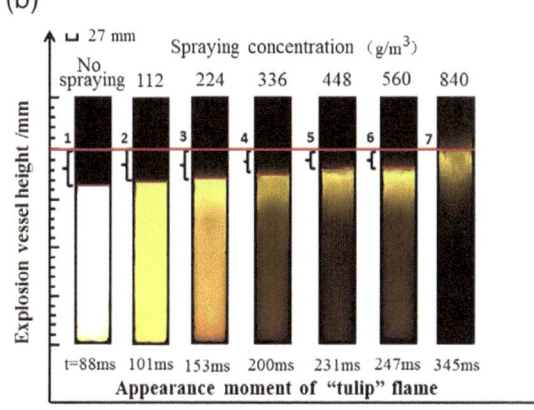

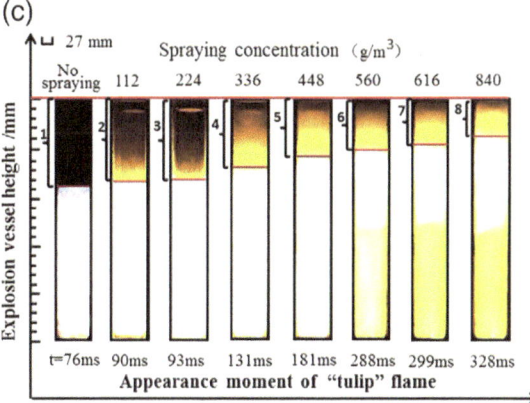

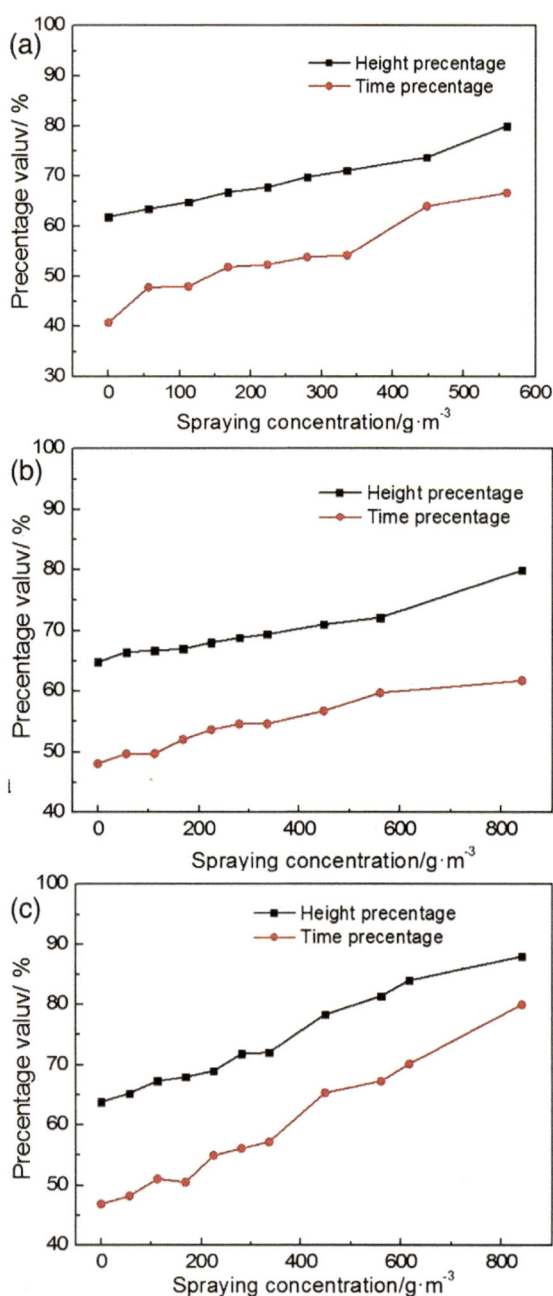

Fig. 8.5 Effect of spraying concentration on the percentage of "tulip" flame appearance. **a** 8% CH₄, **b** 9.5% CH₄, **c** 11% CH₄

large symmetrical vortices behind the flame front. The curvature radius of flame front increased as the symmetrical vortices near sidewalls moved forward and approached the flame front gradually, which would result in the flame surface reduction and "plane" flame appearance. With the enhancement of vortices intensity, the flame propagation velocity near the sidewalls was higher than that in the flame front center zone, resulting in the formation of the "tulip" flame eventually. After spraying mist, the released combustion heat was absorbed obviously in reaction zone and the burned zone. And the explosion intensity and gas combustion rate were weakened obviously with increasing spraying concentration, resulting in the reductions of reverse flow and vortices intensities. As a result, the moving velocities of the vortices near sidewalls and reverse flow in flame front center zone were decreased significantly, which resulted in the delay of "tulip" flame appearance moment. Besides, the moment of "tulip" flame formation was inversely proportional to the laminar burning velocity [5], the reduction of the burning velocity resulted from the heat absorption of water mist would also make the appearance moment of "tulip" flame was delayed obviously.

8.2.3 Flame Flow Field Structure

The addition of ultrafine water mist would cause obvious changes in the explosion flame structure, which means that not only had an effect on the flame reaction zone, but also had a significant impact on the burned zone. After spraying, the flame would appear rhagadiform "cellular" structure. Figure 8.6 presents the evolution process of 8% methane explosion flame "cellular" structure under 336 g/m^3 spraying concentration. The cellular structure underwent four developing stages, "cellular" formation → "cellular" development → "cellular" reduction → "cellular" disappearance. In the initial stage, the flame broke up into large cellular structure (t = 140 ms). With the flame propagating, the large cellular structure at the bottom broke up into small cellular structure, but it was still large cellular structure in the upper zone (t = 210 ms). It means that the flame propagated upward with large cellular structure, and then broke up into small cellular structure gradually (point 2 and 3). The corresponding positions of flame structure variation were marked with number 1–7 in Fig. 8.6. At 235 ms, the cellular structure began to disappear from the bottom to the top of the vessel (t = 270–380 ms). The size of cellular structure closing to the lower flame front was still very small (point 5, t = 270 ms). With the proceeding of flame propagation, the cellular structure was superimposed and transformed into large cellular structure. Eventually, the cellular structure disappeared.

The addition of the mist would cause the formation of cellular structure of explosion flame burned zone. Similar phenomenon was also mentioned by relevant scholars and it was concluded that hydrodynamic instability led to the formation of flame cellular structure [6, 7]. However, the cellular structure would not always appear in methane explosions after spraying mist in our experiments. For 8% methane explosion, the cellular structure would appear after spraying concentration reached 112 g/m^3. And the cellular structure was more significant with the spraying concentration

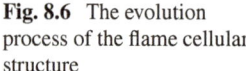
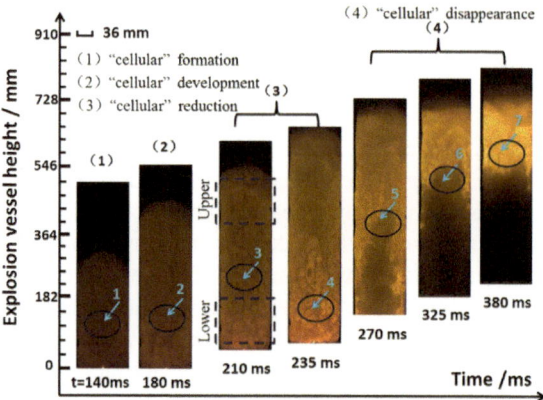

Fig. 8.6 The evolution process of the flame cellular structure

increased. For 9.5% methane explosion, the cellular structure would appear as the spraying concentration exceeded 448 g/m^3. For 11 and 13.5% methane explosions, ultrafine water mist would not cause the appearance of cellular structure; but the explosion flame behind flame front was extinguished acceleratingly with increasing mist concentration. Especially, the 6% and 13.5% explosion flames appeared instability and became irregular as the spraying concentration exceeded 112 g/m^3 and 168 g/m^3, respectively. The cellular structure would not be also observed due to the instability affected by heat absorption of the mist.

The formation of cellular structure could be attributed to the combined effects of hydrodynamic instability and diffusive-thermal instability after spraying mist, which indicated that the mist as a diluent can cause the intrinsic flame instability [8]. Hydrodynamic instability, existing in whole flame propagation process, results from large thermal expansion and can be affected by the flame thickness and flame stretch rate [9]. The diffusive-thermal instability was caused by the mass-heat diffusion disequilibrium and can be characterized by the effective Lewis number. Thinner flame thickness and larger thermal expansion ratio tend to enhance the hydrodynamic instability and the flame thickness and thermal explosion ratio would be decreased after adding mist at atmospheric pressure [10]. The effective Lewis number was reduced slightly as water mist was added into the premixed gas and had a promoting effect on the diffusional-thermal instability. Hence, the appearance of cellular structure was the result of the competition between hydrodynamic and thermal-diffusive instabilities.

The flame radius was small and the flame stretch was large enough to maintain the stable and smooth flame structure in the initial stage of flame propagation after spraying mist. With increasing flame radius, the cellular structure appeared and became more significant due to the enhancement of intrinsic flame instability. Herein, spraying concentration was denoted as Q. As Q < 112 g/m^3 for 8% and Q < 448 g/m^3 for 9.5%, less spraying concentration would not result in the appearance of cellular structure. However, amounts of cracks appeared in the flame surface and showed the tendency of more cellular as the flame propagated upward with increasing spraying concentration. The promoting effects consisted of the diffusional-thermal instability due to the reduction of Lewis number and the hydrodynamic instability due to the reduction of flame thickness, and the inhibiting effect was the hydrodynamic insta-bility due to the reduction of thermal expansion ratio. It indicated that the promoting effects were larger than the inhibiting effect. For 11.0 and 13.5%, the cellular struc-ture would not appear in all propagation stages, however, the flame extinguished acceleratingly in burned zone. It indicated that the inhibiting effect dominated over the promoting effect. According to the occurrence regularity, it is also concluded that the appearance of cellular structure was related to the methane and spraying concen-trations, and water vapor after mist evaporation would exist in the burned zone and continue absorbing heat behind the flame front. The molecular level mixing made water vapor show better heat absorption capacity.

8.3 Effects of Flame Propagation Velocity by Spraying Concentration

Figure 8.7 shows a comparison of the flame front height of 9.5% methane explosion under different spraying concentrations. With the spraying concentration increased, the flame shape showed similar phenomenon. However, the flame front height decreased obviously. The variation of flame front height descended from 106 to 31 mm (t = 40–60 ms). Reduction in changes of flame front height indicated that the flame propagation was obviously inhibited in the initial stage. Moreover, four kinds of flame front shapes were delayed in time successively. As the diameter of the spherical flame increased, the moment of flame touching the inner sidewall was prolonged. And the appearance moment of the "tulip" flame was also delayed from 88 to 231 ms. The flame brightness decreased and changed from blue-golden yellow to dark yellow. The "dark-black" flame in burned zone indicated that the flame was extinguished acceleratingly due to the cooling effect of the mist on the burned products.

Figure 8.8a, b shows the flame height and flame propagation velocity histories of 9.5% methane explosion with the spraying concentration increased. The flame propagation velocity was calculated from the variation of the flame front height with time. As can be seen from Fig. 8.8a, the slope of the flame front height history was decreased and spending time of flame propagation inside the vessel was prolonged obviously with increasing spraying concentration. Figure 8.8b shows that the flame propagation velocity slowed down successively, especially reflected in the reduc-tion of initial stage flame propagation velocity, the decrease of the velocity history slope, the reduction of the Vmax and the delay of its appearance moment. Similar phenomena also existed in the 6.5, 8, 11, and 13.5% methane explosions.

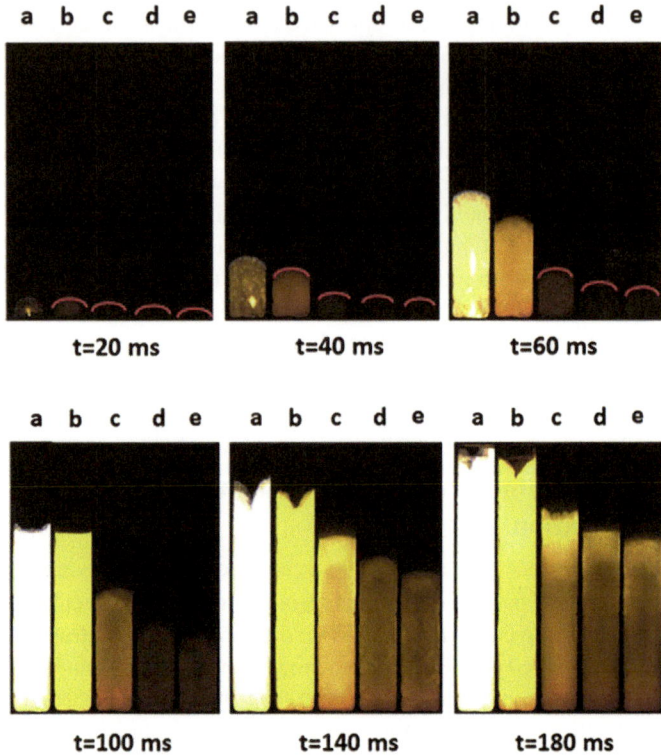

Fig. 8.7 Effect of spraying concentration on flame front height. **a** No spraying, **b** $Q = 112$ g/m^3, **c** $Q = 224$ g/m^3, **d** $Q = 336$ g/m^3, **e** $Q = 448$ g/m^3; 9.5% CH$_4$

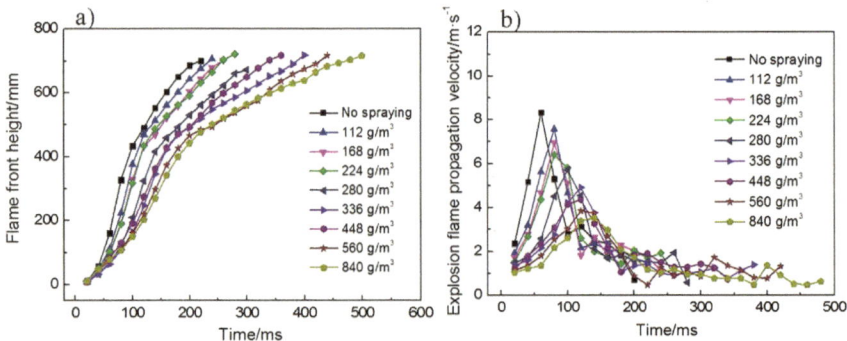

Fig. 8.8 Effects of spraying concentration on flame front height and propagation velocity. **a** Flame front height, **b** flame propagation velocity; 9.5% CH$_4$

8.4   Effects of Overpressure by Ultrafine Water Mist

8.4.1   Overpressure Rising Process

Figure 8.9 shows the relationship between overpressure and pressure rising rate under no spraying (9.5% CH_4). According to the pressure rising rate, the pressure rising process was divided into three stages (I, II, and III in Fig. 8.9). (I) The first accelerating rise stage; (II) the second accelerating rise stage and (III) pressure falling stage. The pressure rising rate history present two peak values (point b′ and d′) and one valley value (point c′), which indicated that the pressure underwent two accelerating rises. The second peak value was greater than first peak value and was called maximum overpressure rising rate ($(dP/dt)_{max}$). At 32 ms, overpressure history began to rise (point a), then the pressure rising rate increased sharply (point a′–b′), and the overpressure history also increased rapidly (point a and b). The first accelerating rise appeared in this stage. After point b, the pressure rising rate decreased obviously. The valley value appeared in point c′, then the pressure history appeared the second accelerating rise process (point c′–d′). The increasing extent of the second accelerating rise was obviously higher than that of the first accelerating rise. The pressure rising rate decreased suddenly after $(dP/dt)_{max}$ (point d′–e′). The ΔP_{max} appeared when the pressure rising rate dropped to zero (point e). After that, the pressure history reduced rapidly.

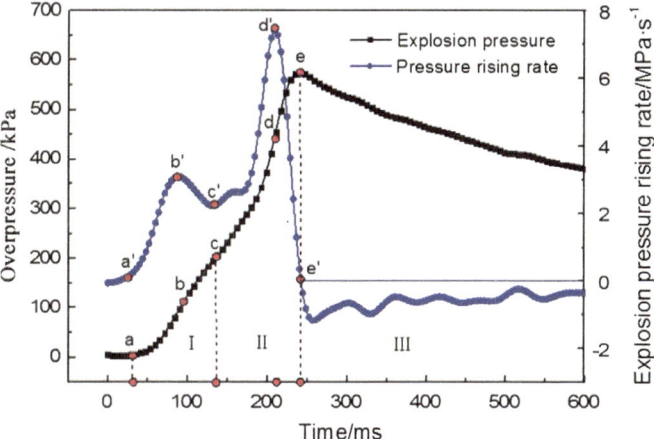

Fig. 8.9   Methane overpressure rising process (9.5% CH_4)

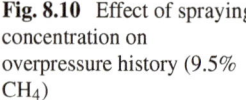

Fig. 8.10 Effect of spraying concentration on overpressure history (9.5% CH_4)

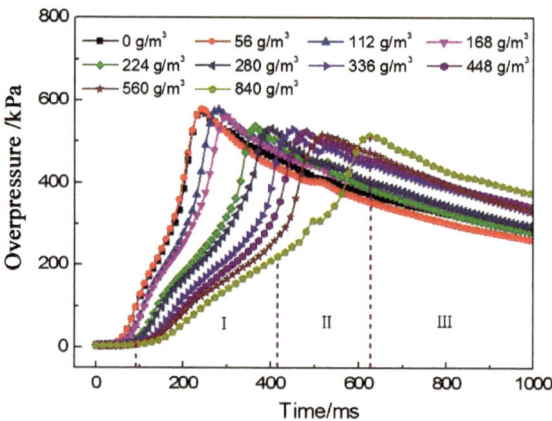

8.4.2 Effects on Explosion Overpressure

Figure 8.10 shows the variation trends of pressure histories of 9.5% methane explosion with the spraying concentration increased. The variation trends of pressure histories with time showed similar characteristics, whereas the pressure rising rate and explosion intensity decreased obviously. With increasing spraying concentration, the moment of overpressure began to rise was delayed from 32 to 95 ms. During the first and second accelerating rise stages, the slope of overpressure history decreased obviously. The ΔP_{max} reduced from 579 to 513 kPa and their appearance moment also was delayed from 239 to 621 ms. Similar phenomena were also observed in 6.5, 8, 11, and 13.5% methane explosions. Table 8.1 shows the ΔP_{max} under different spraying conditions. It indicated that the explosion suppression effect was enhanced with increasing spraying concentration. Especially the flames of 6.5 and 13.5% methane explosions can be inhibited absolutely as the spraying concentration was 336 g/m^3. The flame became weaker and discontinuous above the electrodes and couldn't develop successfully. Meanwhile, the overpressure history would not increase after the ignition.

8.4.3 Effects on Overpressure Rising Rate

Figure 8.11 presents the effect of spraying concentration on the pressure rising rate history (9.5% CH_4). Two peak values and one valley value were presented and the second peak value was higher than the first peak value. Figure 8.12a, b shows the variation trends of two peak values and one valley value and their appearance moments. With increasing spraying concentration, two peak values and one valley value were decreased, and their appearance moments were increased obviously. The first peak value was decreased from 2.82 to 0.95 MPa/s and the corresponding moment was

Table 8.1 The ΔP_{max} and the corresponding moment of methane explosion under different spraying concentrations (unit: kPa and ms)

Methane concentration		Spraying concentration (g/m³)									
		0	56	112	168	224	280	336	448	560	840
6.50%	ΔP_{max}	422	396	397	352	318	233	–	–	–	–
	Δt	1002	1129	1238	1260	1262	1275	–	–	–	–
8%	ΔP_{max}	474	480	458	451	445	432	429	410	405	369
	Δt	392	350	452	544	589	698	750	861	957	1091
9.50%	ΔP_{max}	576	579	578	562	538	531	528	524	510	507
	Δt	230	232	277	294	362	396	447	478	526	626
11%	ΔP_{max}	584	571	570	562	558	555	553	542	514	484
	Δt	217	233	257	278	289	297	330	332	522	716
13.50%	ΔP_{max}	483	466	449	420	371	326	–	–	–	–
	Δt	675	700	907	1114	1330	1542	–	–	–	–

delayed from 85 to 244 ms, and second peak value was decreased from 7.33 to 3.52 MPa/s and the corresponding moment was delayed from 209 to 574 ms. The above result indicated that the first and second accelerating rise processes of overpressure were weakened and the initial and later stages of flame propagation were inhibited obviously. Meanwhile, the valley value was decreased from 1.88 to 0.58 MPa/s, and the corresponding moment was delayed from 134 to 422 ms.

Figure 8.13 presents the reduction degrees of the first and second peak values with the spraying concentration increased compared with no spraying. The reduction degree in first peak value was increased from 8.51 to 66.31%, and the reduction

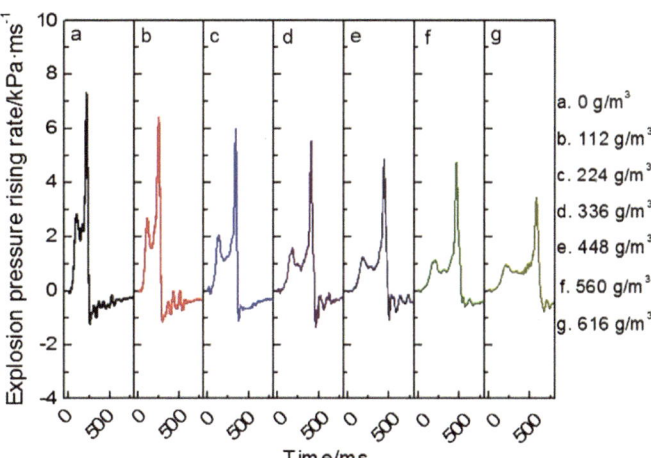

Fig. 8.11 Effect of spraying concentration on overpressure rising rate (9.5% CH₄)

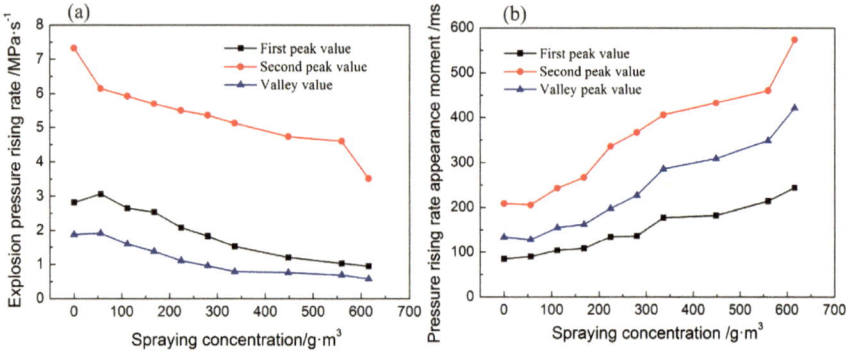

Fig. 8.12 Effects of spraying concentration on two peaks and one valley of overpressure rising rate history and their appearance moments. **a** Peak and valley values, **b** appearance moment

degree of $(dP/dt)_{max}$ was increased from 16.10 to 51.98%. As $Q < 224$ g/m^3, two histories were intersected at point a. As $Q < 224$ g/m^3, the reduction degree of the second peak value was greater than that of the first peak value, which indicated that the suppression effect of the mist on later stage of flame propagation was more significant. Conversely, the reduction degree of the first peak value was greater as $Q > 224$ g/m^3, which indicated that the suppression effect of the mist on initial stage of flame propagation was more obviously. Meanwhile, the time difference of two peak values was enlarged from 124 to 330 ms. It indicates that the effect of explosion suppression was obviously enhanced with increasing spraying concentration.

Reductions in explosion intensity and flame propagation velocity with increasing spraying concentration can be attributed to the enhancements in heat absorbing, heat radiation blocking, gas concentration diluting. First, droplets with size of 10 μm

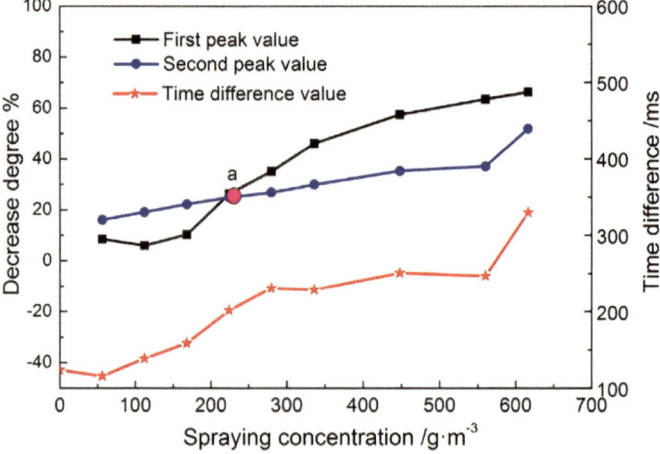

Fig. 8.13 Effect of spraying concentration on the reduction degree of two peak values

would evaporate directly rather than break up by absorbing momentum [11]. The evaporation time scale of droplets 10 μm in diameter was 10^{-2} ms orders [12]. Once touched with high temperature flame front, the mist would absorb amount of combustion heat and evaporate rapidly, resulting in the great reduction of the flame front temperature. Increase in spraying concentration would enhance the capacity of heat absorption for the flame front. According to the Arrhenius formula, reduction in flame front temperature would result in the slowdown of combustion rate.

Second, the water vapor after mist evaporation dispersed widely in the flame reaction zone and unburned zone would dilute methane and oxygen concentration [13]. Meanwhile, the water vapor would also block the heat transfer from combustion zone to unburned zone. With the spraying concentration increased, the effects of gas concentration diluting and heat radiation blocking were enhanced obviously, resulting in the further weakening of heat transfer capacity. The weakening effect would result that the intensity of combustion wave was decreased, and the flame propagation velocity was decreased obviously, resulting in the reduction of pressure rising rate.

Apart from the aforementioned physical effects, the suppression effect of the explosion would also be attributed to the chemical effects. Two chemical suppression mechanisms may be existed as follow. First, water vapor dispersed widely in the flame combustion zone, and the water molecules as a third body would be involved in the chain reaction of explosion process. Amount of high-energy radicals generated from the chain reaction of explosion process would collide with the water molecules. As a result, the radicals would either become inactive with the energy, which were used to activate the branch chain reaction, absorbed by water molecules or react with the water molecules which interrupted the branch chain reaction. Energy absorption of water molecules would weaken the explosion intensity obviously.

Second, water molecules would react with main active radicals (H, O, and OH) of methane explosion process, resulting in the interruption of branch chain reaction. The main participation process can be presented through the following steps. At high temperature, water molecules would stabilize H from atom to molecule ($2H + H_2O \rightarrow H_2 + H_2O$). Due to the effect of the third body, amount of H and OH could be reduced by forming H_2O ($H + OH + M \rightarrow H_2O + M$). Meanwhile, the key reaction steps of producing the main active radicals were also prevented ($HO_2 + CH_4 \rightarrow OH + CH_3O$, $H + O_2 \rightarrow O + OH$). The rate of promoting methane producing (C_2H_6 ($+ M$) $\rightarrow 2CH_3$ ($+ M$)) was greater than that of promoting methane consuming ($CH_3 + O_2 \rightarrow O + CH_3O$) [14]. With spraying concentration increased, the energy of main active radicals and amount of active radicals were reduced obviously. The two chemical suppression processes would decrease the chain reactions rate greatly [15, 16].

Overall, compared with previous suppression experiments, the changes in explosion flame structure and the relationship between pressure rising and flame propagation were analyzed. The formation mechanism of the "tulip" flame was illustrated from the viewpoint of the interactions among the flame front, the flame-induced reverse flow and the vortices and the effect of spraying concentration on the "tulip" flame was researched. The appearance of explosion flame cellular structure was analyzed and its formation resulted from the hydrodynamic instability and diffusive-thermal instability after spraying mist. The change in explosion flame structure as an explanation to reflect the suppression mechanism was investigated. During experiments, without much disturbance were generated in the bulk flow flied to avoid explosion enhancement.

References

1. Xiao, H., Wang, Q., Shen, X., et al. (2014). An experimental study of premixed hydrogen/air flame propagation in a partially open duct. *International Journal of Hydrogen Energy, 39*(11), 6233–6241.
2. Schmidt, E. H. W., Steinicke, H., & Neubert, U. (1953). Flame and schlieren photographs of combustion waves in tubes. *Proceeding of the Royal Society of London A, 4*(1), 658–666.
3. Ponizy, B., Claverie, A., et al. (2014). Tulip flame—The mechanism of flame front inversion. *Combustion and Flame, 161*(12), 3051–3062.
4. Xiao, H., Wang, Q., He, X., et al. (2010). Experimental and numerical study on premixed hydrogen/air flame propagation in a horizontal rectangular closed duct. *International Journal of Hydrogen Energy, 35*(3), 1367–1376.
5. Clanet, C., & Searby, G. (1996). On the "tulip flame" phenomenon. *Combustion and Flame, 105*(1–2), 225–238.
6. Makarov, D. V., & Molkov, V. V. (2004). Modeling and large eddy simulation of deflagration dynamics in a closed vessel. *Combustion, Explosion, and Shock Waves, 40*(2), 136–144.
7. Kessler, D. A., Gamezo, V. N., & Oran, E. S. (2011). Multilevel detonation cell structures in methane-air mixtures. *Proceedings of the Combustion Institute, 33*(2), 2211–2218.
8. Wang, J., Xie, Y., Cai, X., et al. (2016). Effect of H_2O addition on the flame front evolution of syngas spherical propagation flames. *Combustion Science and Technology, 188*(7–9), 1054–1072.
9. Steinberg, A. M., Driscoll, J. F., & Ceccio, S. L. (2008). Measurements of turbulent premixed flame dynamics using cinema stereoscopic PIV. *Experiments in Fluids, 44*(6), 985–999.
10. Kwon, O. C., Rozenchan, G., & Law, C. K. (2002). Cellular instabilities and self-acceleration of outwardly propagating spherical flames. *Proceedings of the Combustion Institute, 29*(2), 1775–1783.
11. Wingerden, K. V., Wilkins, B., Bakken, J., et al. (1995). The influence of water sprays on gas explosions. Part 2: Mitigation. *Journal of Loss Prevention in the Process Industries, 8*(2), 61–70.
12. Adiga, K. C., Willauer, H. D., Ananth, R., et al. (2009). Implications of droplet breakup and formation of ultra fine mist in blast mitigation. *Fire Safety Journal, 44*(3), 363–369.
13. Foersth, M., & Moeller, K. (2013). Enhanced absorption of fire induced heat radiation in liquid droplets. *Fire Safety Journal, 55*, 182–196.
14. Nie, B., Yang, L., Ge, B., et al. (2017). Chemical kinetic characteristics of methane/air mixture explosion and its affecting factors. *Journal of Loss Prevention in the Process Industries, 49*, 675–682.
15. Cao, X., Ren, J., Zhou, Y., et al. (2015). Suppression of methane/air explosion by ultrafine water mist containing sodium chloride additive. *Journal of Hazardous Materials, 285*, 311–318.
16. Liang, Y., & Zeng, W. (2010). Numerical study of the effect of water addition on gas explosion. *Journal of Hazardous Materials, 174*(1–3), 386–392.

Chapter 9
Suppression Mechanism of Ultrafine Water Mist on Confined Gas Explosion

9.1 Numerical Models and Calculation Methods

9.1.1 Physical Model

Gas explosion was a rapid combustion reaction process of flammable gas, and the flame propagates around after ignition. The reaction occurs at the interface between the burned and unburned gases, accompanied by mass and heat transfers. Two prominent characteristics of this process were the sharp increases in the temperature and pressure inside the vessel. Thus, the ultrafine water mist as an inhibitor was adopted to suppress the methane explosion in this study. And the research on gas explosion and its suppression mostly were carried out in a confined space. Such as the study on characteristics of gas explosion and the suppression of gas explosion [1]. Figure 9.1 shows the physical model based on the experimental apparatus. After ignition, the flame propagated as a spherical surface and acceleratingly moved upward after touching the inside wall. The explosion intensity could be effectively reduced due to the heat exchange after coming in contact with the ultrafine water mist distributed uniformly inside the vessel. Furthermore, the vapor pressure after mist vaporization as an important component affected the overpressure inside the vessel and it was also obviously affected by the mist parameters at high temperatures.

9.1.2 Numerical Method

A three-dimensional numerical model for methane explosion suppression by ultrafine water mist was established. Large eddy simulation and partially premixed combustion models were used to determine the explosion flow field characteristics and methane explosion process, respectively. The fluid governing equations of large eddy simulation are obtained by filtering the operation of 3D unsteady conservation equations of

Z. Wang and X. Cao, *Gas Explosion and Its Protection Technology in Process Industries*, https://doi.org/10.1007/978-981-96-3121-6_9

Fig. 9.1 Physical model of
gas–liquid two phases
interaction

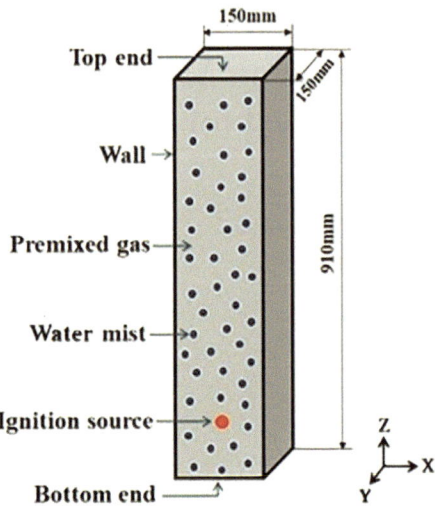

mass, momentum, and energy for a compressible fluid. The Euler–Lagrange equation
was used to solve the continuous and discrete phases, and the coupling calculation
was realized by alternately solving these two phase models.

9.1.3 Boundary Conditions and Calculation Methods

The thicknesses of the vessel end flange and sidewall of a vertical cube structure (910 mm × 150 mm × 150 mm) were 30 mm and 14 mm, respectively. For
vessel walls, non-slip and mass permeability boundary condition were adopted to
the momentum equation ($u = 0$), and zero flux boundary condition was used for
progress variable equation ($\partial c/\partial n = 0$). The heat loss through the vessel walls was
considered, and the outer surface of walls was set to constant temperature boundary
condition of 300 K. The specific heat and thermal conductivity of the vessel walls
were set to 450 J/kg K and 48 W/m K, respectively. The premixed gas was ignited
by the high voltage discharge method at 110 mm above the bottom of the vessel. The
initial pressure and temperature were 0.1 MPa and 300 K. It was hypothesized that
the shape of the ultrafine water mist was spherical, and its surface was smooth, and
the mist did not break up before evaporation. Considering the heat transfer between
the ultrafine water mist and the wall surface, a non-adiabatic boundary condition was
adopted.

   The numerical simulation was carried out using the FLUENT software, which
is based on a control volume method. A coupled solver with explicit linearization
of the equation set was adopted. The time step was determined from the Courant–
Friedrichs–Lewy (CFL) condition. In the process of solving, the transient term was
calculated using the second-order implicit method. The variable equations of the

continuous phase, energy, and reaction process in the convection term were solved by the second-order upwind, and the momentum equation was solved by the central difference method. The variable gradient in the diffusion term was calculated by the least squares cell-based gradient method. Pressure correction was processed using the PRESTO! method, and the PISO algorithm was applied for computing the pressure–velocity coupling process. The coupling calculation was realized by alternately solving the governing equations of the continuous and discrete phases.

9.1.4 Model Validation

The validity of the numerical model was verified by comparing the results of this study with those of our previous experiments. During the experiment, the mist diameter ($d = 10.03$ μm) was measured using a phase Doppler particle analyzer, which ensured the accuracy mist diameter during the mist parameter setting [2]. The generating rate of the mist was measured using a precision balance. Under the action of gravity and drag force, the mist was uniformly distributed inside the vessel after escaping from the atomization apparatus. The mist concentration was determined according to the atomization rate and time, which ensured the consistency of the experiment and simulation working conditions. The pressure change with time and the flame dynamic evolution process were recorded to verify the accuracy of the numerical model.

Figure 9.2 shows the comparison of the overpressure curve and the flame propagation between the experiment and numerical results ($Q_{Mist} = 352$ g/m^3, $d = 10$ μm, and $v = 0$ m/s). As shown in Fig. 9.2, the increase in pressure from the numerical simulation is consistent with that from our experiment. The maximum explosion overpressures (P_{max}) were 516.769 kPa and 515.781 kPa, respectively, and the deviation was 0.19% by calculation. In addition, the maximum deviation of P_{max} is less than 5% by comparing the experimental results under 56, 112, 168, 224, and 280 g/m^3 mist concentrations [3]. Additionally, the flame propagation velocity and flame structure can reflect the development processes of the "spherical", "finger", and "tulip" flames, and have a good consistency in time and space. This indicates that the numerical model and calculation method are accurate.

9.2 Effect of Ultrafine Water Mist on Gas Explosion

The flame front was tracked by solving the reaction progress variable (c). The explosion parameters exhibited a sharp variation in the reaction zone ($c = 0–1$). Figure 9.3 shows the variation curves of the explosion parameters along the axial of the vessel ($Q_{Mist} = 224$ g/m^3; $d = 10$ μm; $v = 0$ m/s; and $t = 90$ ms). It can be seen that the temperature increased sharply in the reaction zone (from 300 to 1372 K), and the pressure presented a similar variation. However, the gas density decreased sharply

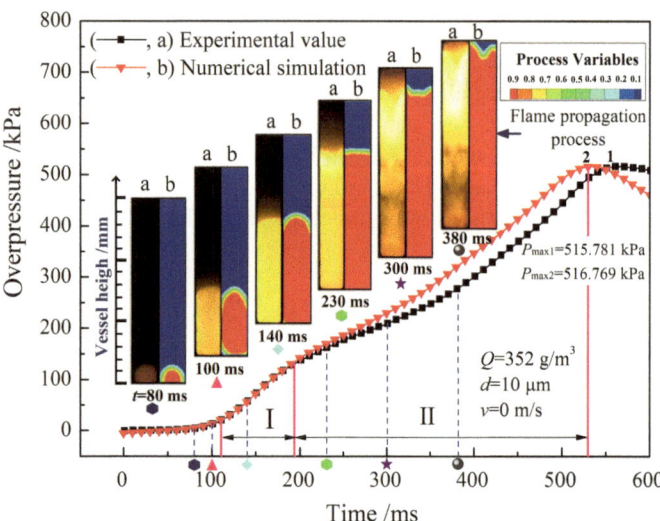

Fig. 9.2 Comparison of the flame propagation and overpressure between the experiment and numerical results

because of the higher temperature. The unburned gas far from the reaction zone was less affected, so the density and temperature are close to the initial condition. The heat cannot be dissipated immediately after gas combustion, which results in a higher temperature and lower density in the burned zone. In addition, the mist was vaporized rapidly in contact with the flame, which results in a dramatic reduction of mist concentration in the reaction zone. As $c = 0.38$ (point 2) and $T = 871$ K (point 1), the corresponding mist concentration decreases to zero. It indicates that the ultrafine water mist was completely vaporized in the reaction zone. This further illustrated that the mist can play an important cooling role on the flame front. Moreover, the ultrafine water mist in front of the reaction zone accumulates due of the action of the compression wave, which results in a higher mist concentration near the flame front than that far from the flame front. The rear of the flame front was the burned zone, and its temperature was related to the endothermic capacity of the mist, which can be affected by the mist parameters. In addition, the heat transfer process between gas and liquid phases during explosion suppression was also obviously affected by the mist parameters.

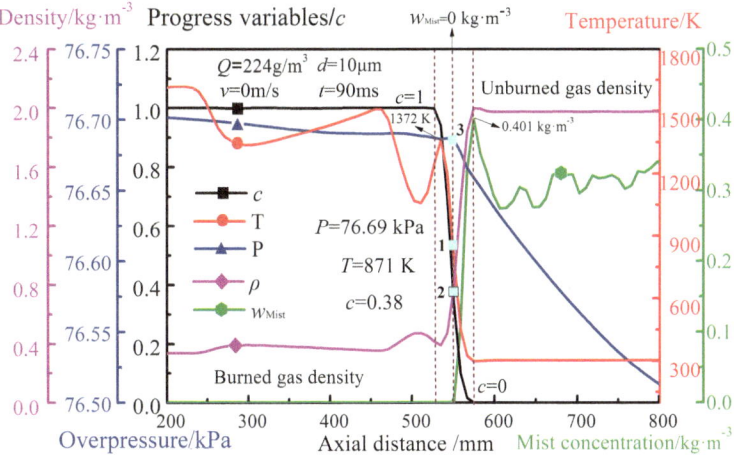

Fig. 9.3 Explosion parameter variation curves along the axial of the vessel (9.5% CH_4, $t = 90$ ms)

9.3 Effect of Mist Parameters on Heat Exchange Rate

9.3.1 Mist Diameter

Figure 9.4 shows the variation curve of the heat exchange rate with increasing mist diameter ($Q_{Mist} = 224$ g/m³; $v = 0$ m/s). P_{max} was an important parameter for estimating explosion intensity. The complete vaporization of water mist before the P_{max} appearance was more conducive to explosion suppression. The abscissa value corresponding to the vertical line was the P_{max} appearance moment. As shown in Fig. 9.4, the moment first increased and then decreased with the increase in mist diameter. The corresponding heat exchange quantity (Q_{Heat}) also presented a similar trend. As $d = 50$ μm, there was still heat exchange occurrence after P_{max} appears. This indicated that the ultrafine water mist was incomplete vaporization at this moment. The larger mist diameter cannot fully play an endothermic role, and the mist amount of incomplete vaporization at P_{max} appearance moment increased continuously with increasing mist diameter, which was not conducive to explosion suppression.

Furthermore, the methane combustion rate was the direct influence factor of temperature rise. The mist diameter indirectly affected the temperature rise rate by influencing the methane combustion rate, as shown in Fig. 9.5. Methane combustion mainly occurred in the reaction zone. As $d = 5$ μm, the mist cannot fully reach the reaction zone to achieve effective endothermic action, which makes that the temperature rise rate was not the smallest under this mist diameter condition. As $d = 10$ μm, the temperature rise rate was the smallest compared with other mist diameter conditions, which indicates that the mist with a diameter of 10 μm has a good contact with the reaction zone and exhibits the strongest endothermic action. With increasing mist diameter (10–200 μm), the temperature rise rate and the maximum explosion

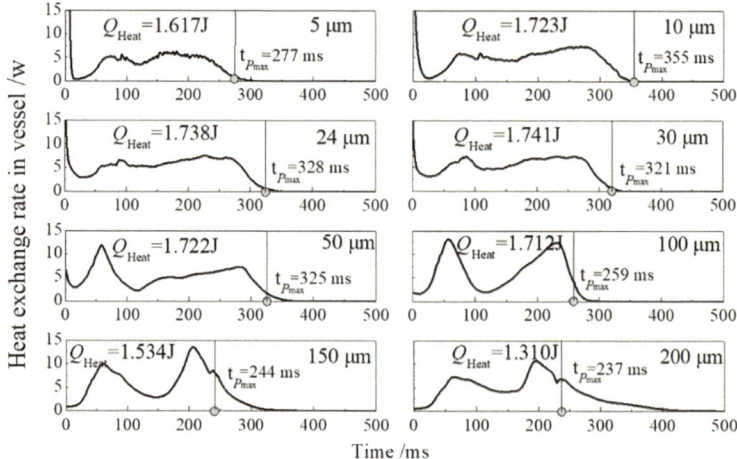

Fig. 9.4 Effect of mist diameter on heat exchange rate inside the vessel

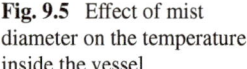

Fig. 9.5 Effect of mist diameter on the temperature inside the vessel

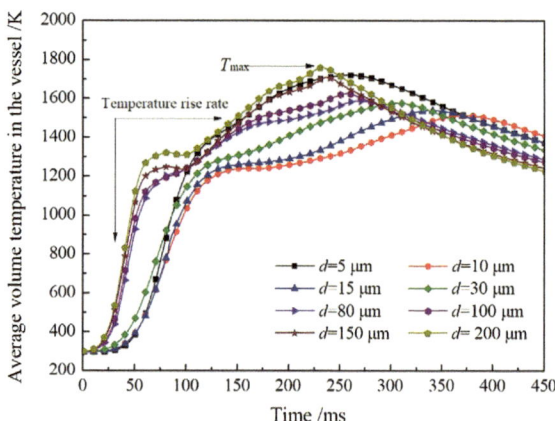

temperature (T_{max}) were increased successively. This indicated that the endothermic capacity of the mist on the explosion flame was decreased.

9.3.2 Mist Velocity

Figure 9.6 shows the variation curve of the heat exchange rate between the gas and liquid phases with the increase in mist velocity ($Q_{Mist} = 224$ g/m³; $d = 50$ μm). As shown in Fig. 9.6, the heat exchange rate obviously increased. This was because an increase in the relative velocity of the two phases can enhance the heat exchange rate between the ultrafine water mist and surrounding gas, which promotes the evaporation of the mist [4]. However, an increase in mist velocity can also accelerate the premix gas combustion and heat release in the reaction zone, except for the enhancement of the heat exchange rate between the gas and liquid phases. This was because the moving ultrafine water mist promotes the mix of the burned and unburned gases and the transport of substances mass [5]. An increase in the heat release rate caused a higher temperature inside the vessel, as shown in Fig. 9.7. It also indicated that the suppression effect of the ultrafine water mist on the gas explosion was reduced with an increase in the mist velocity.

Fig. 9.6 Effect of mist velocity on heat exchange rate inside the vessel

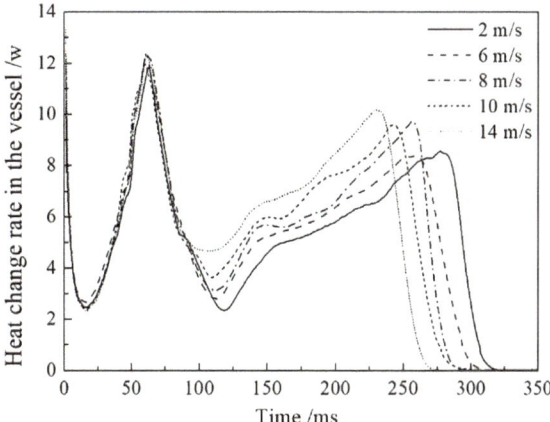

Fig. 9.7 Effect of mist velocity on the temperature inside the vessel

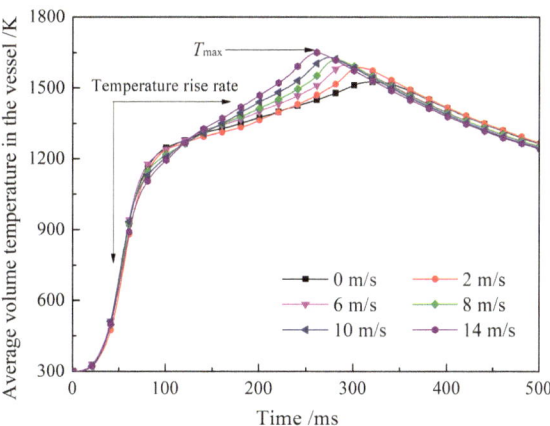

9.3.3 Mist Concentration

Figure 9.8 shows the variation curve of the heat exchange rate between the gas and liquid phases with increasing mist concentration ($d = 10$ μm; $v = 0$ m/s).

As shown in Fig. 9.8, the heat exchange rate was the smallest under the 112 g/m³ mist concentration condition. With increasing mist concentration (112–562 g/m³), the heat exchange rate can be increased significantly. However, the curve slope decreased successively. Especially for 674 g/m³ mist concentration, the extent of reduction was the most significant because of the reduction of explosion flame temperature resulting from the enhancement of the heat absorption capacity of the mist. The combined effects of mist concentration and explosion flame temperature affected the heat exchange rate and exchange time between the two phases. The integral of the heat exchange rate with time showed that the heat exchange quantity can be increased significantly (from 0.8619 to 5.1436 J). This indicated that the endothermic capacity was enhanced, and the explosion reaction rate was reduced with increasing mist concentration.

Figure 9.9 shows the effect of mist concentration on the temperature inside the vessel under the corresponding mist condition. It can be seen that the temperature decreased in turn with increasing mist concentration. This was because of the enhancement of the heat absorption capacity by the mist. Furthermore, the temperature in the reaction zone was reduced after coming in contact with the mist, which results in a decrease in the heat release rate. Thus, the temperature inside the vessel showed a decreasing trend because of the obvious reduction in the heat accumulation rate.

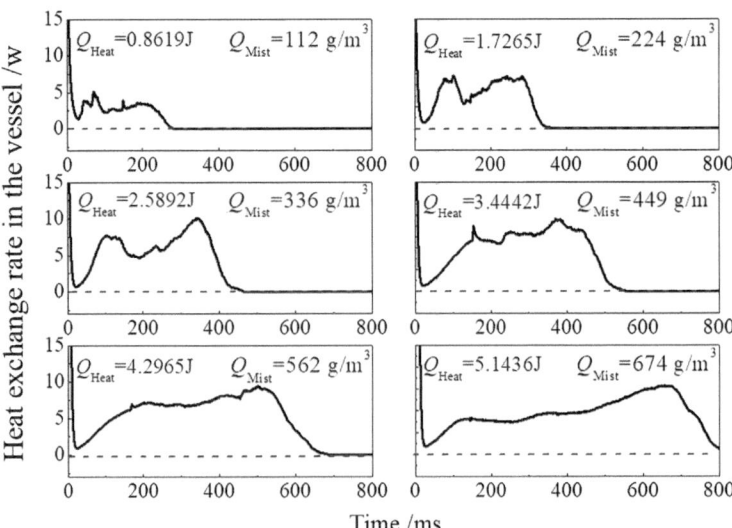

Fig. 9.8 Effect of mist concentration on heat exchange rate inside the vessel

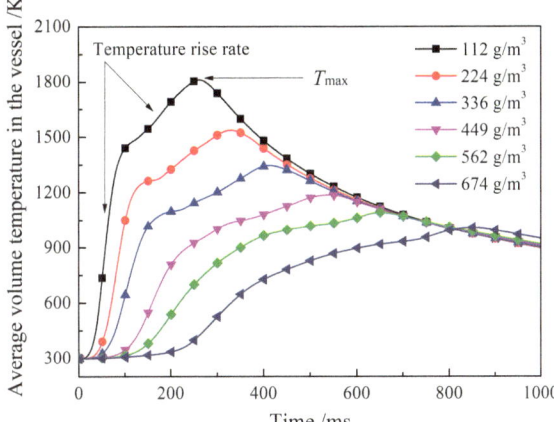

Fig. 9.9 Effect of mist concentration on the temperature inside the vessel

9.4 Effect of Mist Parameters on Vapor Pressure

Heat released by methane explosion results in a sharp increase in the temperature inside the vessel. At high temperatures, apart from the pressure generated from gas components (combustion products and unburned gases), the vapor generated from the mist vaporization can also produce a larger vapor pressure, which further affects the overpressure inside the vessel. Figure 9.10 shows the corresponding relationship between the overpressure, ultrafine water mist vaporization rate, vapor pressure, and the ratio of vapor pressure to overpressure with time in the closed vessel (Q_{Mist} = 224 g/m^3; $d = 10$ μm; and $v = 0$ m/s). The vapor pressure was calculated according to the amount of mist vaporization and the corresponding temperature. As shown in Fig. 9.10, the vapor pressure increases with the explosion time. The vaporization rate curve undergoes two accelerating rise processes [6], and the corresponding vapor pressure curve also presents a similar rise trend. This indicates that vapor pressure was directly affected by the water vaporization rate, and their rise processes were consistent. By calculation, the vapor pressure accounts for 27% of the overpressure. This illustrates that vapor pressure was an important component of overpressure in a closed vessel. This was the result of the combined effects of the mist vaporization rate and temperature. The mist parameters can indirectly affect the vapor pressure by influencing the mist vaporization rate and temperature.

9.4.1 Mist Diameter

Figure 9.11 shows the variation curve of the ultrafine water mist vaporization rate with increasing mist diameter. The abscissa value corresponding to the vertical line in the figure was the P_{max} appearance moment under different mist diameter conditions. It can be seen that the mist vaporization rate first increases and then decreases. The

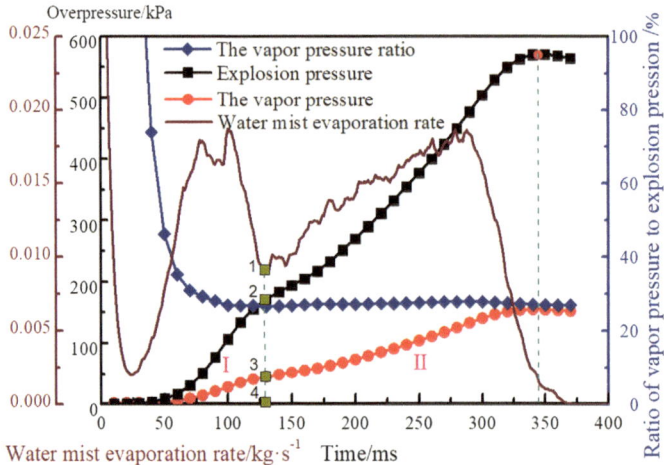

Fig. 9.10 Relationship between the mist vaporization rate, the vapor pressure, and the ratio of vapor pressure to overpressure inside the vessel

P_{max} appearance moment also presents a similar variation trend. As $d = 50$ μm, vaporization still occurs after P_{max}. It also indicates that the ultrafine water mist inside the vessel was incomplete vaporization at this moment as the mist diameter exceeds a certain value, and the incomplete vaporized mist amount increases continuously with increasing mist diameter.

Figure 9.12 shows the effect of mist diameter on vapor pressure under the corresponding mist conditions. As shown in Fig. 9.12, the vapor pressure was the least

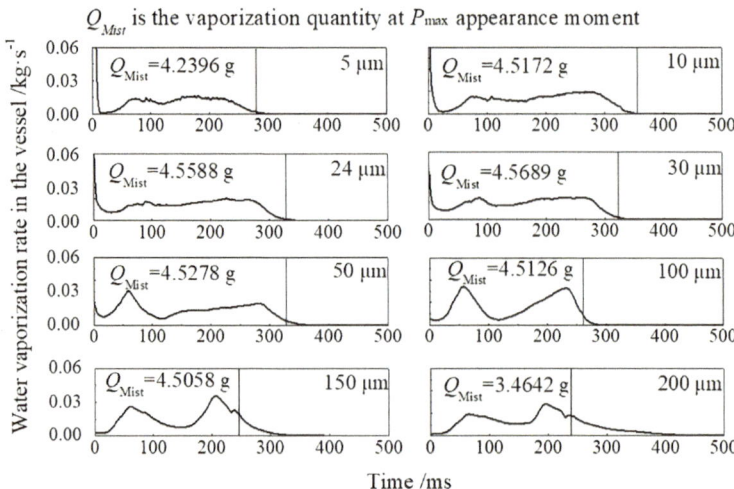

Fig. 9.11 Effect of mist diameter on the mist vaporization rate

Fig. 9.12 Effect of mist diameter on vapor pressure

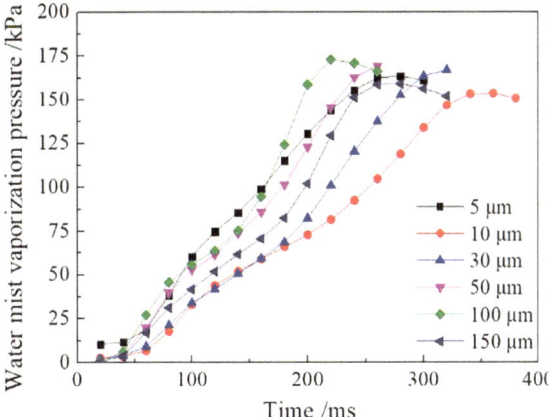

when the mist diameter was 10 μm. With increasing mist diameter ($d = 10$–100 μm), the vapor pressure presents increasing trend. Especially for a mist diameter of 100 μm, the maximum vapor pressure appeared and then decreased with increasing mist diameter. The mist diameter indirectly affected the vapor pressure by affecting the mist vaporization rate and temperature. As $d = 5$ μm, the higher temperature inside the vessel resulted in a larger vapor pressure (in Fig. 9.5). As $d = 10$ μm, the ultrafine water mist had better contact with the flame front and realizes a good heat exchange, resulting in obvious reductions in the explosion reaction rate and temperature. Thus, the vapor pressure and its rise rate decreased due to the combined effect of the above two factors. With an increase in the mist diameter (10 μm $< d < 100$ μm), the temperature inside the vessel was increased continuously, which resulted in an increase in the vapor pressure and its rise rate. As $d > 100$ μm, the mist vaporization rate and its vaporization amount decreased significantly. Although the temperature was increased, the vapor pressure presents a decreasing trend under the combined effect of the above two factors.

9.4.2 Mist Velocity

Figure 9.13 shows the variation curve of the ultrafine water mist vaporization rate with increasing mist velocity ($Q_{Mist} = 224$ g/m^3; $d = 50$ μm). As shown in Fig. 9.13, the mist vaporization rate obviously increases. This was because that an increase in mist velocity can enhance the heat exchange rate between the gas and liquid phases. In addition, the turbulent intensity of the flame front can be also enhanced, which results in an increase in the explosion reaction rate significantly. Thus, the mist vaporization rate increases due to the above two influencing factors. Figure 9.14 shows the effect of mist velocity on vapor pressure under the corresponding mist condition. As shown in Fig. 9.14, the vapor pressure increases successively with

Fig. 9.13 Effect of mist velocity on the mist vaporization rate

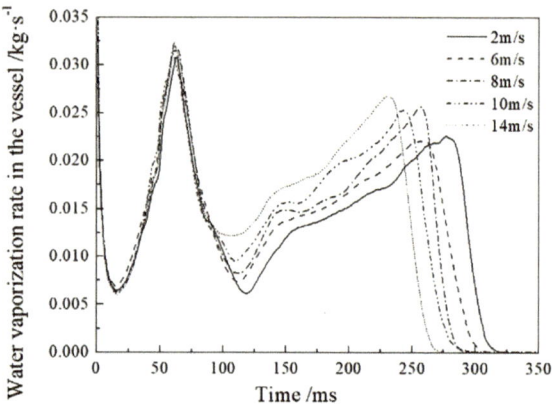

Fig. 9.14 Effect of mist velocity on vapor pressure

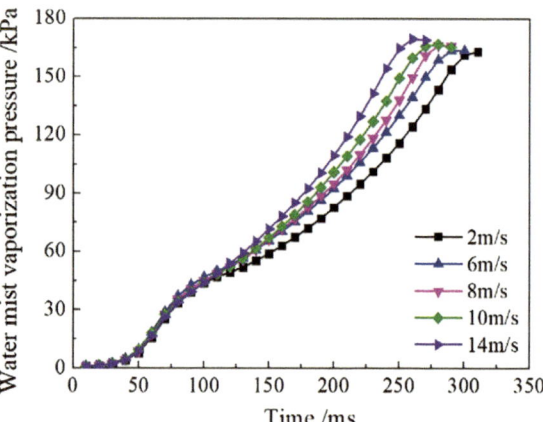

increasing mist velocity. This was because of the simultaneous increase in the mist vaporization rate and temperature inside the vessel. Thus, the relative velocity of the gas and liquid phases should be decreased as much as possible to achieve effective explosion suppression.

9.4.3 Mist Concentration

Figure 9.15 shows the variation curve of the ultrafine water mist vaporization rate with the increasing mist concentration ($v = 0$ m/s; $d = 10$ μm). As $Q_{Mist} = 112$ g/m^3, the mist vaporization rate inside the vessel and the time required for mist complete vaporization are the smallest, as shown in Fig. 9.15. The mist vaporization rate increased with the increasing mist concentration (224–562 g/m^3). However, the curve slope decreases, and the time required for mist complete vaporization is significantly

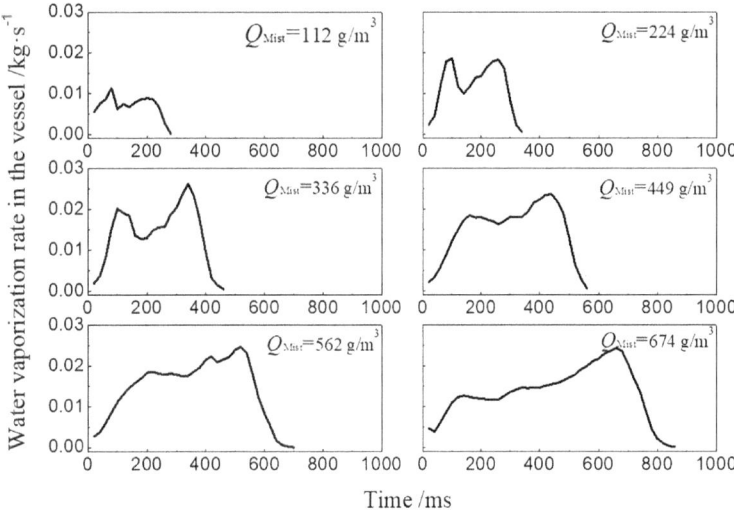

Fig. 9.15 Effect of mist concentration on the mist vaporization rate

prolonged. In particular, for 674 g/m³ mist concentration, the reduction extent of the mist vaporization rate is the most significant, and the time required for mist complete vaporization is prolonged from 282 to 860 ms. This indicated that increasing the mist concentration can enhance the endothermic capacity and decrease the methane explosion reaction rate and temperature inside the vessel.

Figure 9.16 shows the effect of mist concentration on vapor pressure under the corresponding mist conditions. The vapor pressure inside the vessel increased as the increase in mist concentration; however, the rise rate decreases because of the enhancement of heat absorption capacity, as shown in Fig. 9.16. As $Q_{Mist} = 112$ g/m³, the temperature inside the vessel is higher (in Fig. 9.9); however, the amount of mist vaporization is lower, which results in a lower vapor pressure due to the combined effect of the above two factors. As $Q_{Mist} = 674$ g/m³, the temperature inside the vessel is lower; however, the vapor pressure is larger due to the greater amount of mist vaporization. The vapor pressure rising rate is smaller due to the lower mist vaporization rate. Thus, it was concluded that the mist concentration should be increased as much as possible to effectively suppress gas explosions.

Fig. 9.16 Effect of mist concentration on vapor pressure

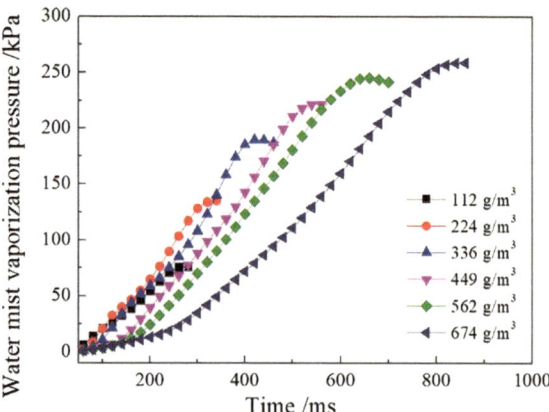

References

1. Zheng, K., Yang, X., Yu, M., et al. (2019). Effect of N_2 and CO_2 on explosion behavior of syngas/air mixtures in a closed duct. *International Journal of Hydrogen Energy, 44*(51), 28044–28055.
2. Cao, X., Ren, J., Bi, M., et al. (2017). Experimental research on the characteristics of methane/air explosion affected by ultrafine water mist. *Journal of Hazardous Materials, 324*, 489–497.
3. Cao, X., Ren, J., Bi, M., et al. (2016). Experimental research on methane/air explosion inhibition using ultrafine water mist containing additive. *Journal of Loss Prevention in the Process Industries, 43*, 352–360.
4. Ragland, K. W., Bryden, K. M., & Kong, S. C. (2011). *Combustion engineering.* CRC Press.
5. Yu, M. G., An, A. N., & You, H. (2011). Experimental study on inhibiting the gas explosion by water spray in tube. *Journal of China Coal Society, 36*(3), 417–422.
6. Cao, X., Bi, M., Ren, J., et al. (2019). Experimental research on explosion suppression affected by ultrafine water mist containing different additives. *Journal of Hazardous Materials, 368*, 613–620.